HANDBOOK OF

MICROBIOLOGICAL CRIT

CW01668753

Institute of Food Science & Technology
5 Cambridge Court
210 Shepherd's Bush Road
London W6 7NJ

ACKNOWLEDGEMENTS

Working Group Members (1999):
Chris Bell – Convenor
Melody Greenwood
Jane Hooker
Alec Kyriakides
Rebecca Mills

Working Group Members, I (2009):
Peter McClure – Convener
David Alexander
Chris Bell
Alec Kyriakides
Christine Little

Working Group Members II (2018):
Peter McClure – Convener
John Bassett
Roy Betts
Ching Chan
Marianne James
Intisar Khan
Clare Menezes
Phil Voysey
Andy Walker

Grateful thanks are due to many colleagues who gave generously of their time in contributing to and commenting on various drafts of this document. Thank you also to Kiu Sum, Mariam Zaki and Rachel Ward for acting as IFST Secretariat Editors.

PREFACE

The first edition of this book was published in 1997 and an update was issued in 1999. The purpose of this revision, long overdue, is to include European Commission legislation (EC No. 2073/2005) that came into force on 1st January 2006 and to include other advances enabling the document to reflect current understanding and concepts relevant to microbiological criteria and their application for food quality and safety.

Food microbiology has seen some truly remarkable developments in the past 30-40 years, involving innovative technologies in food processing and packaging, development and adoption of robust food safety programmes, new methods for recovery, isolation and characterisation of foodborne microorganisms, new concepts and tools for assessing risk and predicting the fate of microorganisms in foods, and a better understanding of the mechanisms underlying the response of microorganisms to their environment. Despite these advances, foodborne disease remains a significant threat to public health, driven by failures and mistakes in the preparation, manufacturing and handling of foods, very often involving poor hygienic practices and non-adherence to some basic principles that support preventative programmes such as HACCP.

Considering this continuing threat to public health, the intended purpose of this document remains unchanged i.e. to provide practical working guidance for those involved in producing, using and interpreting microbiological criteria in the food and catering industries. Therefore, the aims are still valid: the publication brings together microbiological criteria derived from the practical experience of the authors, and existing guidelines and standards.

Written by professional food microbiologists with wide experience, it discusses definitions, derivation and limitations of microbiological criteria, and sets out tables for different commodities and technologies. We have also updated the document to reflect recently developed microbiological methods, changes to taxonomy, inclusion of recently emerged pathogens and a brief description of recently developed processing technologies.

I would like to acknowledge the hard work, dedication and perseverance of the two working groups who spent long hours working to enable IFST to produce this version of what has proved to be a very useful guide for the food industry. We remain strong believers in the quality assurance philosophy and hope this guide will help to promote this more widely.

Peter McClure
Chair and Convener of IFST Microbiological Criteria Working Group

CONTENTS

1. INTRODUCTION

1.1. Aims and Scope

The safety of food is best assured using a preventative approach based on the application of food safety management systems such as HACCP (Hazard Analysis and Critical Control Point), at all stages in the food supply system, supported by good manufacturing and hygienic practices, and other pre-requisites.

It is recognised that microbiological criteria are widely used in the food industry, but they have rarely been published. Whilst not advocating the introduction of further legal standards or the indiscriminate use of criteria, food industry experience of practical microbiological criteria needs to be shared.

This document discusses the role of microbiological testing and criteria in support of a safe food supply. It aims to provide guidance to all those involved in producing, using and interpreting microbiological criteria in the food and catering industries. It has been written by professional food microbiologists with experience in food manufacturing and retailing, public health, food examination and government in consultation with colleagues who are also expert in these fields.

Microbiological criteria are inevitably controversial, and the figures given in the tables (see Section 6) are derived from the practical experience of the authors as well as industry best practice, existing specifications, guidelines and standards. Such a document can never be exhaustive, and the food categories used encompass a wide range of products. Where microbiological testing and criteria are considered useful by a food business, the information and values given in this document may be used as a basis for determining relevant criteria for individual products.

Criteria should be applied not only to finished products but also ingredients, process intermediates and the manufacturing environment. For example: to provide assurance that food is manufactured under hygienic and sanitary conditions, an environmental monitoring programme should also be included within the site testing plan. This should include indicators and/or pathogens related to the risk associated with the product. This will give a good insight into the base level contamination risk within the manufacturing site, to the exposed finished product post heat-treatment (or another relevant kill-step).

This guideline does not cover the details of design, development, deployment and implementation of an Environment Monitoring Plan. Further details on how to establish this can be found in but not limited to Control of *Salmonella* in Low-Moisture Foods (GMA, 2009) and similar governmental guidance documents.

It is hoped that this document will continue to stimulate a productive discussion on this important subject.

Note: The values in the tables have been developed by expert elicitation and are intended to provide guidance in this area and have no official status.

1.2. Background

Foodborne disease and microbial spoilage of food is the result of the failure or inability to control microorganisms at one or more stages of the food chain, from raw material production to consumption of the final product. The implications of situations that result in food poisoning outbreaks or food spoilage can be severe for consumers, food producers, retailers and regulatory authorities.

Traditionally, control of microorganisms in a food has been demonstrated by microbiological testing of samples at various stages of production and of the final product. Criteria have been developed to give some degree of assurance that food is safe and of suitable quality, and that it will remain so to the end of its shelf-life provided it is handled appropriately. Conformance to microbiological criteria is one tool that is most commonly verified by testing e.g. Certificate of Analysis (CoA), but other tools such as Certificate of Conformance (CoC) may be used. It is recognised that microbiological testing can never give an absolute assurance of product safety or quality, particularly because of the problems involved in product sampling and the distribution of microorganisms in food. Used judiciously however, microbiological testing is a valuable tool in the food industry.

In the food industry today, microbiological testing can be effective when used alongside physical and chemical process control systems to support hazard analysis approaches such as HACCP (ICMSF, 1988; Codex, 2003) and other HACCP-based systems. The HACCP system is a structured approach to identifying hazards and defining and implementing systems of control. Microbiological examination can be used to help verify that these critical control points and pre-requisite programmes remain under control.

Modern approaches to food safety management often now include consideration of a variety of risk-based microbiological targets such as Food Safety Objectives (FSOs), Performance Objectives (POs) and Performance Criteria (PCs) (Codex, 2005).

A microbiological **Food Safety Objective** is defined as the maximum frequency and/or concentration of a microbiological hazard in a food considered tolerable for consumer protection, i.e. one that provides or contributes to the Appropriate Level of (health) Protection (ALOP).

A **Performance Objective** is the maximum frequency and/or concentration of a hazard in a food at a specific step in the food chain before consumption that provides or contributes to an FSO or ALOP.

The **Performance Criterion** is the outcome of control measures at one or more steps in the food chain and at these steps, hazards can be prevented, eliminated or reduced. Examples include:

- 12D reduction of proteolytic *Clostridium botulinum* in low acid canned foods;
- 6D reduction of *Listeria monocytogenes* in ready-to-eat chilled foods;
- 6D reduction of psychrotrophic strains of *Clostridium botulinum* in pre-prepared; refrigerated foods with extended shelf-life;
- 5D reduction of *Escherichia coli* O157:H7 for fermented meat products.

Note: The 'D value', also called 'decimal reduction time', is the time required to destroy 90% or 1 log10 reduction of the spores or vegetative cells, for specific microorganism, in a specific substrate at a specific temperature (Campden BRI, 2006).

These measures, often referred to as metrics, are used to indicate the level of control at a specific step in a food safety risk management system and microbiological criteria can be considered as a metric in a similar way. The main purpose of FSOs and POs is to translate the ALOP into levels of hazards in the food chain that can be communicated and managed by the industry. As recommended by EFSA (2007), it is important that the goal for risk management is established before evaluating possible control options. The application of microbiological criteria at different steps in the food chain, including primary production, food processing environments, processing lines and finished product is provided in ICMSF texts (ICMSF, 2011).

1.3. Definition and Practical Application of Microbiological Criteria

A microbiological criterion consists of, as a minimum, the following elements:

- The microorganism, microbial toxin or metabolite of concern;
- The food concerned and sample type;
- The sampling plan (see 3.3.1);
- The microbiological limit (s).

The other aspects that should be included in a criterion are purpose, specified point in food chain where criterion applies, reason for selecting microorganism (in addition to the microorganism), indication of statistical performance of sampling plan, analytical methods, performance parameters, and actions that should be considered in the event of the criteria being breached.

Conformance to microbiological criteria should be verified using specified, relevant methods which have been validated for the microorganism or toxin of concern in the food being examined (see also 4.4 and 6.1).

Microbiological criteria are essentially of three types:

- **Standard**

- **Guideline**
- **Specification**

These terms, as defined by the Codex Alimentarius Commission (Codex, 1981), were principally for use in Codex standards and codes of practice. Microbiological standards, guidelines and specifications have been redefined by Codex (Codex, 2013a), the International Commission on Microbiological Specifications for Foods (ICMSF, 1986) and the United States Subcommittee on Microbiological Criteria (National Research Council, 1985) and for simplicity the following working definitions are generally recognised:

Standard

This is a microbiological criterion contained in a law or a regulation where compliance is mandatory. These are introduced by governments or regulatory authorities. Typical examples include most criteria in European Community (EC)/European Union (EU) Directives and Regulations, and the Statutory Instruments of England, Wales, Northern Ireland and Scotland. The food industry must ensure full compliance with these standards which are monitored by enforcement agencies or the industry themselves. Products not complying with standards will be removed from the supply chain as unfit for the intended use (human consumption as per EU regulation) if based on food safety criteria. Products already on sale must be withdrawn and, if necessary, recalled from the consumer. If process hygiene criteria are not met, the food business operator must investigate and take relevant corrective action. Current EU and UK microbiological standards are summarised in Section 7, including Regulation (EC) No. 2073/2005 (and amendments).

Guideline

A microbiological guideline is a criterion applied at any stage of food processing and retailing which indicates the microbiological condition of the sample and aids in identifying situations requiring attention for food safety or quality reasons. Results obtained from testing against a microbiological guideline also assist in trend analysis. Results that deviate significantly from the trend may indicate a tendency towards a situation which is out of control and highlight the need for attention before control is lost. Guidelines are usually self-imposed by the food industry but may occasionally be included in legislation. Products not complying with guidelines should result in investigative action to identify and rectify the cause and consider impact on the product. Guidelines on the levels and types of microorganism relevant in specified foods produced under good manufacturing practice may also be provided by industrial associations for their members. In the UK, the Health Protection Agency (HPA) – now Public Health England (PHE) - has published microbiological guidelines to assist food examiners and enforcement officers in the determination of the microbiological quality of foods and to indicate levels of bacterial contamination considered to be a potential health risk in ready-to-eat foods at point of sale (HPA, 2009).

Specification

This is a microbiological criterion, developed by the food industry, applied to raw materials, ingredients or the end-product which is used in a purchase agreement. Criteria may include pathogens, toxins, indicator microorganisms or spoilage microorganisms where non-compliance may affect product safety and/or quality during shelf-life. Industry product specifications have historically been more stringent than microbiological standards to provide a margin of safety but recent strengthening of legislative standards e.g. EC Regulation No. 2073/2005 on Microbiological Criteria for Foodstuffs, have made such differences less pronounced for the microorganisms included in such standards.

Products not complying with specifications should be investigated to determine the cause. Rejection of products may occur even if they are not hazardous or unwholesome at the time of testing.

It is important to ensure that the components of microbiological specifications are relevant and realistic and fully understood by both parties in the agreement.

1.4. Components of Microbiological Criteria

Microorganism or Microbial Toxin of Concern

The contaminants detailed in the criterion may be foodborne pathogens, food spoilage microorganisms, indicator microorganisms or microbial/biological toxins (ICMSF, 1978). Specified microorganisms must be relevant to the food and/or food process (see Appendix I Specific microorganisms and the principles of test methods). Pathogen testing is applied for reasons of public health where evidence exists of a potential hazard to health from a specific food/microorganism. Testing for indicator and spoilage microorganisms is used for the following reasons:

Trend analysis: Indicator and spoilage microorganisms are usually present in higher numbers than pathogens and indicate unsatisfactory conditions when levels increase significantly. Since levels can be monitored, trends can be established so that trend analysis can identify situations before they become out of control. Trend analysis may also be useful for monitoring the incidence of pathogens in raw foods by assessing their frequency of occurrence and level of contamination. Trend analysis can be used on any microbiological data set.

- **Cost-benefit:** More analyses can be conducted for microbial indicators of unhygienic practice and spoilage microorganisms than for specific pathogens because testing for pathogens is usually more expensive. However, this consideration should not supersede pathogen testing where it is deemed more beneficial to public health to test for pathogens

- **Simplicity:** Tests for indicator microorganisms are frequently simpler to conduct than tests for specific pathogens.

- **Rapidity:** Tests for indicator microorganisms usually provide results more rapidly than those for pathogens, thus allowing for faster remedial action.

Food Concerned and Sample Type

Microbiological criteria must indicate the food and stage of processing to which they relate. Ideally, they should give a description of the food, indicating key processing features and conditions under which the material should be stored and used. Such factors significantly influence the content and validity of a microbiological criterion. For example, a specification for frozen food may allow higher aerobic colony counts at the point of production compared with the same chilled food since bacteria will not grow under frozen conditions but may grow during chilled storage. However, the growth of bacteria during or after defrosting should be considered for frozen products.

Also, the microbial levels in a perishable sample tested at the beginning of its shelf-life are likely to differ from those at the end of shelf-life. Such differences must be considered when specifying a microbiological criterion and it should be made clear at what point in the product shelf-life the relevant criterion is to be applied. It is also important to indicate the nature of the sample if this is likely to influence the end result and subsequent interpretation. Microbiologically sensitive components of multi-component foods e.g. sandwiches may need to be tested separately. In the food industry such details are usually contained in product specifications.

Microbiological Limit

The microbiological limit defined in a criterion represents the level above which action is required. Levels should be determined from knowledge of the microbiology of raw materials and the effects of processing, product handling, storage and end-use (including foreseeable misuse) of the final product. It should be recognised that some environmentally-sourced, unprocessed raw materials e.g. flour, may contain pathogenic microorganisms.

Limits for pathogens, spoilage microorganisms and indicators may be derived by determining microbial levels in products where conditions of processing, storage and product characteristics are known. They should consider the levels that indicate inadequate control and may represent a hazard to health.

Some microbiological limits in foodstuffs may vary by season and geographical source. Consequently, there is a need to consider the microbiological levels that are achievable as well as desirable. In some instances, for example for hygiene criteria, it may be more appropriate to set 'targets' rather than true limits.

It is not possible to prove the absence of a microorganism in a product and negative results from detection (presence/absence) tests should be declared as 'not detected' (ND) e.g. not detected in 25g or not detected in 1ml. It is important to note however, that where the criterion specifies ND in 25g e.g. for a pathogen in a ready-to-eat food, it should not be interpreted that presence in a higher quantity is acceptable e.g. in 50g.

Presence of pathogens where the level is specified as ND in 25g may represent an unacceptable food safety risk if present at even lower levels e.g. presence of *Salmonella* species in any sample weight of a ready-to-eat food would not be deemed acceptable.

For quantitative tests in which no growth occurs, it is normal to record the result as less than the limit of detection of the test e.g. <10 cfu/g, which is equivalent to ND in 0.1g.

1.5. LIMITATIONS OF MICROBIOLOGICAL TESTING

The tests used to assess whether a food complies with a microbiological criterion are subject to several factors that should always be clearly understood when using them.

Methods

No microbiological method can detect all representatives of the target microorganism being sought. The presence of sub-lethally injured cells or cells that may be viable but non-culturable, contaminants present in the food but at a level below the sensitivity of the method and variants not capable of growth under the conditions of the test may result in failure or sub-optimal recovery of the target microorganism. Confidence limits for results from enumerative techniques are also relatively wide (BSI, 2015) (See section 4).

DNA testing methods such as PCR (polymerase chain reaction) and WGS (whole genome sequencing) are becoming increasingly available for detecting and characterising pathogens and can have several advantages over standard microbiological culture techniques. The technique of PCR detects specific markers (DNA sequences) present in an organism and can be used as a screen of an enrichment culture.

Whole genome sequencing is a method for determining the complete DNA sequence of an organism, which can be used for both detection and detailed characterisation. This technique has several uses, particularly when it is important to know if the same organism is present in multiple samples. This may include outbreak investigation of foodborne illness, or during source attribution (e.g. an investigation to determine whether the same organism has been found on multiple occasions across a plant to try and find its source).

When using such techniques, it is essential to note that a positive result can be obtained from the DNA of non-viable cells even at low levels as very low copy numbers of DNA can be detected. Therefore, it is essential to ensure that cultural methods are also used initially to confirm the viability of the target organism and as such these techniques are to be considered as an addition to conventional cultural methods and not a replacement.

Sampling

Microorganisms are rarely distributed homogeneously in a sample or a product batch and pathogens, if present, are usually at low levels. Any single test for detection or enumeration only reflects the situation in that particular sub-sample of the batch.

The only way microbiological testing and associated criteria can be totally reliable is by adoption of 100% inspection with methods of 100% reliability and accuracy. Since current methods are effectively destructive, 100% testing would leave no food to consume and whilst this would undoubtedly reduce foodborne illness it is not a feasible option.

Sampling is therefore based on the application of statistical techniques; the degree of assurance derived from a result will depend in part on the number and size of samples taken. Detection of foodborne pathogens, present at low levels and distributed heterogeneously in a food, may require an exceptionally high number of samples to be analysed to achieve a reasonable degree of assurance (ICMSF, 2018).

The validity of the test result is also dependent on sample size and type and sampling techniques; aseptic sampling procedures are of particular importance for preventing extraneous contamination.

Laboratory Competence

Laboratory staff competence and the operation of quality systems in laboratories are very important for achieving valid microbiological results. Independent auditing, the use of appropriate external quality assurance (QA) schemes and accreditation of laboratories all help to provide assurance of a laboratory's competence (Wood and Corry, 2003; Campden BRI, 2013).

International Standards Organization (ISO) standard 17025 defined the minimum requirements for the QA system as applied to specific tests, in addition to the management structure of the laboratory (ISO, 2005).

Effectiveness

Despite the limitations described (which are well understood by food microbiologists), correct use of microbiological testing plays an important part in support of hazard analysis and control (ILSI, 2004). When used regularly, non-statistical sampling and testing schemes (see section 3) can provide a simple form of monitoring (trend analysis) that can allow trends to be identified and add to the assurance that the process is under control. This constructive use of microbiology, alongside other process control systems helps to ensure that the food supply is safe and of good quality.

1.6. HACCP

HACCP and HACCP-based systems are recognised throughout the world as being the most effective way of identifying hazards to food processes and products and identifying means by which the hazards may be controlled, and the control systems monitored (Codex, 2003).

Microbiological testing alone cannot prove the safe design or production of foods and should never be relied on for this purpose. However, microbiological testing can be used to support an effective HACCP programme adding to the existing assurances that process controls have achieved the desired effect. For example, a raw material containing Enterobacteriaceae that is subject to a full heat process.

A critical control point (CCP), capable of achieving their destruction may be tested after the process to demonstrate that these microorganisms were destroyed i.e., they are not detected by the test method used. This microbiological test may be used as an additional verification check relating to the efficacy of the process; however, it should only be used in conjunction with the more practical CCP process checks of temperature and heating times. Monitoring temperature and time will provide relevant in-line information that is more useful as any faults identified can be recognised and corrected immediately. Many processors seek additional assurance by testing foods for specified microbial targets for due diligence purposes and HACCP verification.

Microbiological testing should never be carried out at the expense of a HACCP-based system of controls but may be used as a useful adjunct to both HACCP and risk management.

2. FACTORS AFFECTING THE MICROBIOLOGY OF FOODS

2.1. General

The food industry uses a wide variety of technologies to produce an enormous range of foodstuffs, from chilled foods with short shelf-lives to shelf-stable intermediate moisture foods and long-life packaged foods. Each technology involves a complex series of interlinked processes that combine to produce safe and wholesome products. It is essential to understand the interrelationships of these processes so that relevant criteria can be developed and applied. Monitoring procedures should be derived from HACCP-based assessments of all the processes involved.

Factory structure, hygienic design, manufacturing equipment, raw materials, product formulations and packaging systems, process flows, temperature/time controls and hygiene systems (personnel and environment) all affect the microbiological status of the finished product and each requires careful evaluation.

2.2. Factory structure, environment and personnel

Unless processed within their final sealed pack, with a step capable of eliminating microorganisms present, all products will be susceptible to contamination from the processing environment from sources such as air, water, plant equipment, dust and personnel. Some of these contaminants may be tolerated whereas others may impact the safety or quality of the foodstuff and as such need to be controlled. Where such controls are in place to minimise the risk of contamination it is important to both validate and verify their effectiveness. Therefore, it is as important to develop appropriate criteria and sampling plans for the environment as it is for the final product.

In fact, environmental testing often provides far more assurance than finished product testing as the whole production run passes through that environment as opposed to testing just a small amount of a batch where non-homogeneous contamination is a significant limiting factor. The distribution and degree of contamination from the environment itself is determined by the level of exposure of the foodstuff, as well as the variety of microbes within that environment, thus even here the result is often sporadic and highly variable. For example, if testing final product where the source of contamination is the food handler, testing the food handler themselves is likely to identify the problem whereas testing the final product or swabbing the plant depends on contact having been made with the surface/product sampled. The impact of such variation can be reduced by understanding the likely routes of contamination and targeting these as part of a risk-based sampling programme.

When considering the appropriate criteria, firstly the relevant organisms must be selected. This will relate largely to those selected in the finished product specification but must also consider organisms for which this specification is not appropriate. For example, toxigenic species may grow within the process and render the product unsafe through production of heat stable toxins whereas the actual bacteria may be eliminated by a heat step. This would therefore not give rise to a problem being identified at the finished product test point, whereas process sampling where such growth can occur is where the assurance testing is relevant e.g. testing for coagulase positive staphylococci in slow process vat tanks for cheeses.

It is therefore important to fully understand the product, process and have sufficient knowledge of food microbiology to determine the microbes of concern and how they may contaminate the foodstuffs as they pass through the process.

As a general guide, the following may be considered subject to specific risk assessments:

(a) Indicator organisms – implies a potential toward loss of control but does not identify food safety concerns by themselves. Levels need to be set according to what indicates acceptable vs unacceptable and are often unique to the plant and environment:

- Enumeration of Enterobacteriaceae – useful indicator for hygiene standards in non-raw post process areas;

- Enumeration of yeasts and moulds – useful indicator of post process air quality and plant hygiene;
- Detection of *Listeria* species – useful indicator in post process chilled environments for plant hygiene.

(b) Food safety organisms – identifies a potential or confirmed threat to product safety. Levels are defined by food safety requirements and are usually consistent across all plants and environments:

- Detection of *L. monocytogenes*;
- Detection of *Salmonella*;
- Enumeration of coagulase-positive staphylococci;
- Enumeration of *Bacillus cereus*.

Often, a combination of both indicator and pathogen testing is appropriate, although frequency of testing and sampling points may differ.

'What to sample' is determined by risk of contamination to the foodstuff. A systematic approach is required to identify potential contamination vectors from start of process to finish. This must consider existing process controls in place and actual risk to the final product. For example, it would not be appropriate to sample for *Salmonella* in a raw process area if there is adequate control designed at a later point to both eliminate such organisms and segregate post elimination to prevent contamination from raw to final product, whereas a sampling plan to verify the effectiveness of the segregation control should be in place by sampling the post process area. Alternatively, in processing environments where there is no final kill step, environmental sampling may be a key verification activity.

'When to sample' should consider both pre and post clean environments and plant as the testing needs to provide assurance of different factors, which are all important to regularly assess:

- Pre-start-up sampling of food contact surfaces;
- Pre-clean sampling, to verify the maintenance of adequate control during manufacture and is best performed immediately prior to cleaning as this also verifies appropriate runtimes and cleaning schedules;
- Post-clean sampling, to verify the cleaning standard.

'Sample type' is determined by what will give the most information from the test. Main sample types for environment and process assessment are:

- Swabs/sponges, which are useful to assess both plant/equipment and environmental surfaces but as in common with food testing the homogeneity of the contamination is a limiting factor;
- Settle plates or air samplers may be useful to assess air quality and movement of airborne contaminants throughout a production area;
- Rinse samples, that are useful to assess cleaning standards particularly of cleaning in place (CIP) systems;
- Foodstuff samples, that are very useful to assess impact of contamination at a particular point in the process and may consist of intermediate product, plant residues, finished product;
- Hand swabs can be a useful indicator of post process personnel hygiene both post wash and whilst manufacturing. However, results should be used to educate personnel and drive standards rather than as a means to screen out personnel.

Usually a combination of the above is used to provide a very effective means of verifying the hygiene standard of the plant and environment.

'Sampling and Test frequency' is determined by the risk of loss of control and impact on the product. This can vary from daily for higher risk products through to weekly or monthly sampling. In addition to this routine schedule there may be requirement for additional testing for plant validation or in the event of issues or following maintenance/construction events. Rotation of sampling points may also be appropriate since conditions may change.

Other important steps include data evaluation, implementation of a corrective action plan to respond to findings and review of the sampling programme. A risk-based well managed environmental testing programme can provide much greater assurance and aid in investigations than finished product testing alone.

The IFST publication "Food and Drink - Good Manufacturing Practice: A Guide to its Responsible Management" (IFST, 2018) contains useful advice concerning premises, equipment and approaches to good manufacturing practice (GMP).

2.3. Raw materials

It is essential to understand the nature of potential hazards that may be introduced by raw materials and the potential for these to affect finished product. Specifications for raw materials should contain relevant microbiological, toxicological and physico-chemical parameters.

Microbiological criteria for finished product should take account of the level and types of microorganism present in a raw material, those that may come from the processing environment (including from food business employees) and, the processing conditions that will affect the levels of microorganisms throughout the process. Even where a manufacturing process may be expected to eliminate a particular potential microbial hazard from the raw material, the possible production by the microorganism of heat-stable toxins which may cause illness, or enzymes (proteases and lipases) that may cause spoilage; all need to be controlled at the raw material stage with appropriate microbiological criteria applied at that point. This includes water used as an ingredient and water that is used in processing e.g. in jacketed vessels, that may come into contact with intermediate or finished product.

2.4. Process Technology

General

A clear understanding of the effect of food process technologies on microorganisms and their survival and growth characteristics in foods is necessary before relevant criteria can be assigned (ICMSF, 1996). It is sometimes the case that products contain microorganisms that are of no consequence to the safety or quality of that product, considering its shelf-life and preparation before consumption and therefore, setting criteria for these microorganisms is pointless e.g. presence of low levels of thermophiles in canned products destined for temperate climates. Indeed, some microorganisms are an inherent part of the process/food e.g. in cheese, yoghurt.

Temperature

Foods may be ambient, chilled, frozen, mildly heat treated (thermised, pasteurised, sous vide) or exposed to high heat treatments (Ultra High Temperature (UHT) or those used in some drying or canning processes).

Freezing to temperatures below approximately -10°C will stop microbiological growth but enzyme activity can still occur and may cause organoleptic changes. The microbial population at the point of freezing generally remains stable throughout the frozen shelf-life although some microorganisms are more susceptible to freezing temperatures and may reduce in number.

Chill temperatures prevent the growth of many microorganisms, but some pathogens and a wide variety of food spoilage microorganisms may grow, albeit usually more slowly. Criteria applied at the point of manufacture must take account of the potential for subsequent growth of psychrotrophic microorganisms during shelf-life.

Mild heat treatments such as pasteurisation reduce the levels of vegetative microorganisms in foods but thermotolerant organisms, and spores may survive, and some toxins will remain biologically active. These processes are often used in conjunction with other factors, such as storage at chill or reduced pH or low water activity (a_w) to extend the shelf-life of products. Criteria must take account of these microorganisms together with post-process contaminants.

High temperature processes such as UHT and some canning processes are used to produce shelf-stable products. Vegetative microorganisms, including most pathogenic species, should be completely destroyed and only the spores of a few types of spore-forming bacteria are able to survive such processes.

Water Activity

Water activity (a_w) is a measure of the water available for microbial growth (i.e. water not bound to other molecules); reducing a_w in a food can restrict the growth of many microorganisms. Water activity may be reduced from the maximum (1.0 for pure water) to low levels (<0.6) by direct removal of water e.g. drying, or by the addition of solutes such as salt or sugar.

The microorganisms capable of growth or survival at the a_w of a particular food should be considered when setting criteria. For example, most bacteria cannot grow below a_w 0.9 but *Staphylococcus aureus* can grow down to approximately a_w 0.85 and may be important in criteria for some dried products, whilst some yeasts and moulds can grow at levels as low as a_w 0.6-0.65.

pH and Type of Acid

The pH and levels of organic acid of a product affects the range of microorganisms that can survive and grow. Many products have a low pH i.e. < pH 4.5 resulting from natural acidity, fermentation or pickling processes. Spoilage microorganisms included in criteria for such products are likely to be yeasts, moulds and acid tolerant bacteria. Pathogenic bacteria may survive, but few are likely to grow. At a product pH greater than 4.5, *Clostridium botulinum* may grow and produce toxin. It may therefore be necessary to apply a high heat process (sterilisation process, F_0 >3.0 see section 5.5) or acidify the product to ensure its safety. The possibility of outgrowth of microorganisms that may survive any heat treatment and raise the product pH to a level at which pathogens may grow should also be taken into account.

An additional consideration is the type of acid used, where the effects of organic acids are dependent on pH and dissociation constant (pK_a) of the acids used. Such information helps to identify the microorganisms of potential concern, some of which will be selected for microbiological criteria.

Atmosphere

The atmosphere within product packaging is often altered to help improve product keeping quality and increase shelf-life. Air may be removed altogether to create a vacuum as with some meat joints, or air may be replaced by other gases e.g. nitrogen and/or carbon dioxide in various proportions as with many sliced cooked meats, cheese, fresh salad mixes and fish products. Most of these latter types are termed modified atmosphere packs (MAP) and the atmosphere within these packs changes over time because of the dynamic processes that occur between the food and the atmosphere, and because of the permeability of some packaging materials.

Strictly controlled atmosphere packs (CAP) are those in which these dynamic processes are balanced throughout the storage life of the product, the atmosphere therefore remaining constant. Some novel packaging systems incorporate chemical-based scavenging systems to keep oxygen levels low. In each case, the packaging material is carefully selected to ensure its permeability characteristics are appropriate for the required atmospheric conditions to be maintained. Such packaging systems are often used for chilled foods and although the changed atmospheric conditions contribute to the inhibition of some microbiological growth, lactic acid bacteria and yeasts in particular may cause spoilage.

Consideration must be given to the potential for some modified atmospheres to provide conditions which select for the growth of clostridia including non-proteolytic *C. botulinum* (ACMSF, 1992). Furthermore, there are processes that could lead to conditions supporting growth of clostridia e.g. reduction in redox potential caused by growth of other microbiota.

Combinations of Factors

The combined effect (often referred to as hurdle technology) of non-optimal conditions of temperature, water activity, salt concentration, pH and packaging atmosphere on the growth and survival of microorganisms usually has a greater effect than an individual factor used at the same level (ICMSF, 1996). Preservatives can also be used in combination with these factors to enhance microbial suppression and inactivation.

There are a number of texts that provide extensive information on each of the above factors in addition to listing some 'limits for growth' of some microorganisms for water activity and pH values. Many of the combinations that may be used to control microorganisms in foods are listed in ICMSF book 5 (ICMSF, 1996).

It is important to understand the interactions of the various combinations and other factors that might impact on their controlling activity. For example, chill storage conditions will inhibit or slow the growth of many microorganisms but will allow survival for long periods whilst storage at higher temperatures that still prevent growth often promote reductions of certain contaminating microorganisms e.g. Enterobacteriaceae. These considerations are particularly important for products that rely on more traditional preservation systems, such as fermentation.

Developing Technologies

A variety of non-thermal technologies such as irradiation, high pressure processing (HPP, or ultra-high pressure, UHP), ultra-filtration, pulsed light and pulsed electric fields (PEF) have been, or are being, investigated for application in specific food market sectors. All such technologies aim to reduce microbial loads and vary in their efficacy and potential areas of application. If required, microbiological criteria in relation to such foods can be developed from a consideration of all the factors discussed in this guideline. It is important to understand the mode of action of these technologies and the response of various target microorganisms so that meaningful criteria are set. Apart from thermal sterilisation, only irradiation can be used in isolation to produce ambient-stable low acid foods. As with thermal processing, non-thermal technologies may be used in combination with other factors so that robust preservation systems are established, and shelf-life can be extended. Examples of such combinations include HPP with low pH (e.g. orange juice) or reduced water activity (jams) or both (fermented meats), and HPP with storage at refrigeration temperatures (guacamole). These combinations take advantage of the lethal effects of high pressure on vegetative pathogens and the inhibitory effects of other factors on those microorganisms (e.g. spore-formers) able to survive the high-pressure treatment.

The effect of the technologies used and subsequent product handling, including product storage and distribution on conditions as well as consumer handling, all contribute to the safety and wholesomeness of the food. Account must be taken of all these stages for the determination of relevant microbiological criteria.

Shelf-life

The potential for growth of and/or toxin production by residual microbial populations in finished products depends on the types of microorganisms present and their ability to grow to a level of concern under the storage conditions applied during the product shelf-life. Not all organisms need to grow to present a hazard. As such limiting the shelf life of the food will not diminish the food safety risk.

Microbiological criteria should take account of any microorganisms likely to be present in the finished product and their relevance to the safety, quality and shelf-life of a product. The levels of tolerance applied at the point of manufacture should be such that allowing for predictable growth of these microorganisms, the product will remain safe and wholesome to the end of shelf-life provided it is stored under the correct conditions. It is industry best practice to test beyond shelf-life (e.g. 1.1-1.2 x shelf-life), to ensure robustness. This allows for foreseeable misuse of product.

2.5. APPLICATION OF MICROBIOLOGICAL TESTS

Microbiological testing, even using rapid methods, takes time (sometimes many days) and costs money. There is a tendency to include too many microorganisms and to apply over-stringent limits in microbiological criteria in a misguided attempt to achieve product safety. Inappropriate criteria are costly in time and resources (human and financial) which could be put to better use in the practical application of HACCP and GMP systems. It is therefore important to ensure that any such testing is both relevant and meaningful within the context of the food production operation. An understanding of the ecology and biology of target microorganisms is essential, in conjunction with an appreciation of the statistical aspects of sampling and testing (ICMSF, 2018). Answering the following questions will assist in obtaining the most sensible, practical and cost-effective use of microbiological testing:

- What information is needed?
- Why is this information necessary?
- What is the best means of obtaining the information?
- If a microbiological test is to be used, at what stage is it best applied taking account of the variation that occurs?
- What criteria are appropriate and practicable?
- What response is appropriate to 'above limit' results?

For example, tens of thousands of finished product examinations for *Staph. aureus* are carried out each year in industry on extensively handled foods, and the microorganism is rarely detected. To minimise the level of *Staph. aureus* in the finished product, it may be more effective, in these circumstances, to apply a structured personnel hand-swabbing and training programme for those involved in direct handling of the product and/or its components. This application of the test could yield more positive benefits in heightened personnel awareness and improved personal hygiene practices and handling procedures than would be accrued from so many tests on product.

Product safety and due diligence obligations will be supported more cost effectively by the application of directly relevant microbiological testing regimes and criteria based on sound product and process knowledge in association with HACCP-based systems and the operation of GMP.

3. SAMPLING FOR MICROBIOLOGICAL TESTING

3.1. INTRODUCTION

No amount of testing will ever make a food safe. The sheer volume and variety of food produced, the distribution and state of microorganisms in the food, and industry resource limitations make it essential to ensure that food production operations are inherently safe. Microbiological testing should be regarded as a helpful tool in support of these systems, not as a means whereby a 'satisfactory' result from an end product is used to demonstrate the safety of the food.

The reliability of a test result from the analysis of a food is dependent upon the extent of sampling, types of sample taken, the quality of the aseptic sampling procedures used, and the accuracy and precision of the method employed.

Inappropriate sampling can lead to invalid conclusions being reached. The purpose and limitations of sampling and the relevance of statistical techniques to sampling need to be understood if the results from microbiological tests are to be correctly interpreted.

3.2. TERMS

The term 'sample' refers to the material removed for examination from an identified batch or lot. A batch (or lot) of product is defined in different ways by different individuals according to their specific requirements and may vary e.g. from a single mix of ingredients to a continual production run over a defined time period (hours or days). A sample may consist of one or more sample units. The individual sample unit may be split into one or more test portions, or sample units may be combined for testing. The test portion will usually be only part of the sample unit and it is essential to understand that the microorganisms in a sample unit are unlikely to be distributed evenly so the test result will relate strictly, only to the test portion quantity examined. This underlines the importance of ensuring that sampling is as representative as possible at each sub-sampling step. For the following discussion, unless otherwise indicated, no distinction is made between sample, sample unit or test portion.

When sampling plans are compared and their reliability or stringency in making decisions is considered, different aspects of their performance should be considered. No practical sampling plan can achieve perfect discrimination between acceptable and unacceptable lots because it would require all samples in the lot to be tested to identify an unacceptable lot. A full discussion of statistical sampling and methods is given in ICMSF texts (ICMSF, 2018).

The criteria and sampling plans given by the ICMSF are generally recommended for port-of-entry sampling and examination of foods i.e. circumstances in which very little may be known concerning the food production processes (see 2.1). Food manufacturers, however, should have full knowledge of the raw materials, sources and specifications, processing technologies and other factors contributing to the finished product's microbiological characteristics. With this knowledge, sampling and microbiological testing in food businesses can be applied in a more practical way (see 1.2).

The size, number, composition and method of taking samples will affect the outcome of a test (test result) and should be specified. Attention to consistency with respect to these aspects of sampling will maximise the repeatability and reliability of the test result, but uneven microbial distribution may still cause results to vary significantly between tests.

The sampling plan should also take account of the nature of the hazard and the likelihood of it occurring given the raw materials, the process and process controls. Hazard categories and associated sampling plans have been detailed (ICMSF, 2018). Historical data demonstrating the absence of a hazard are often used in determining a sampling plan but, in the absence of a proper hazard analysis approach assumptions based on such data are unsound.

3.3. SELECTION OF SAMPLING PLAN

Sampling Plan Structure

The sampling plan should describe the sampling strategy to be used in assessing the microbial target and indicate the action to be taken in response to test results. The principles of 2-class and 3-class attribute plans are described by ICMSF (2002) and these are commonly applied in industry.

Sampling plan definitions may be summarised as follows:

Two-class sampling plan – results are either satisfactory or unsatisfactory

n = number of samples analysed;

m = maximum acceptable level of target microorganism(s) or toxin(s);

c = the number of samples allowed to exceed m.

In industry, 2-class sampling plans are applied mainly when testing for specific pathogens using detection (presence/absence) tests and c=0.

Example: In a two-class sampling plan for *Salmonella* in a ready-meal, where 5 samples each of 25g are taken and tested and any positive test results are unacceptable, the following applies:

n = 5, c=0, m = not detected in 25g

This is a simple test protocol. A specified number of samples and quantity (weight) are examined. If a 'not detected' result in all these samples is achieved, the test requirements are satisfied.

Obviously, there is a greater chance of detecting the target organism when larger quantities of test portions are examined e.g. 100g rather than 10g. Also, as microorganisms are usually unevenly distributed in foods, greater confidence in their detection may be achieved by specifying larger numbers of samples for examination, and by sampling different portions of the food, e.g. core vs edge. Table 1 illustrates this approach.

Table 1: Probability of accepting a batch containing specific % levels of defective product

Number of sample units examined	Probability of Acceptance of a Defective Batch			
	Actual % of defective samples			
	10%	**20%**	**30%**	**40%**
3	73	51	34	22
5	59	33	17	8
10	35	11	3	1
20	12	1	<0.5	<0.5

(Modified from ICMSF, 2018)

The figures given are the percentage probability of obtaining negative results and accepting a batch containing 10-40% actual defective product (positives). For example, in a two-class plan where c=0, if five samples are tested and the true incidence of defective sample units is 30%, then the chance of accepting the batch is 17% even though negative results are obtained on all five samples tested.

If 10 samples are tested and found to be negative, then for the same incidence of defectives, the chance of accepting the batch is 3%, i.e. the larger the number of samples examined, the smaller the probability (chance) of accepting a batch at any specific true incidence of defective product.

Such information is calculated and often presented in 'Operating Characteristic' curves which readily demonstrate the probability of accepting or rejecting batches in relation to the true incidence of defective product (ICMSF, 2018). An example of this is shown in Figure 1.

Figure 1: The operating characteristic curve for a sampling plan with n = 10, c = 2

(ICMSF, 2018)

i.e., the probability of accepting lots in relation to the proportion defective among the sample units comprised in the lot being examined. If, for example, the lot comprises 20% (p=0.2) defective sample units, the lot will be accepted with a probability of 0.68 and rejected with a probability of 0.32.

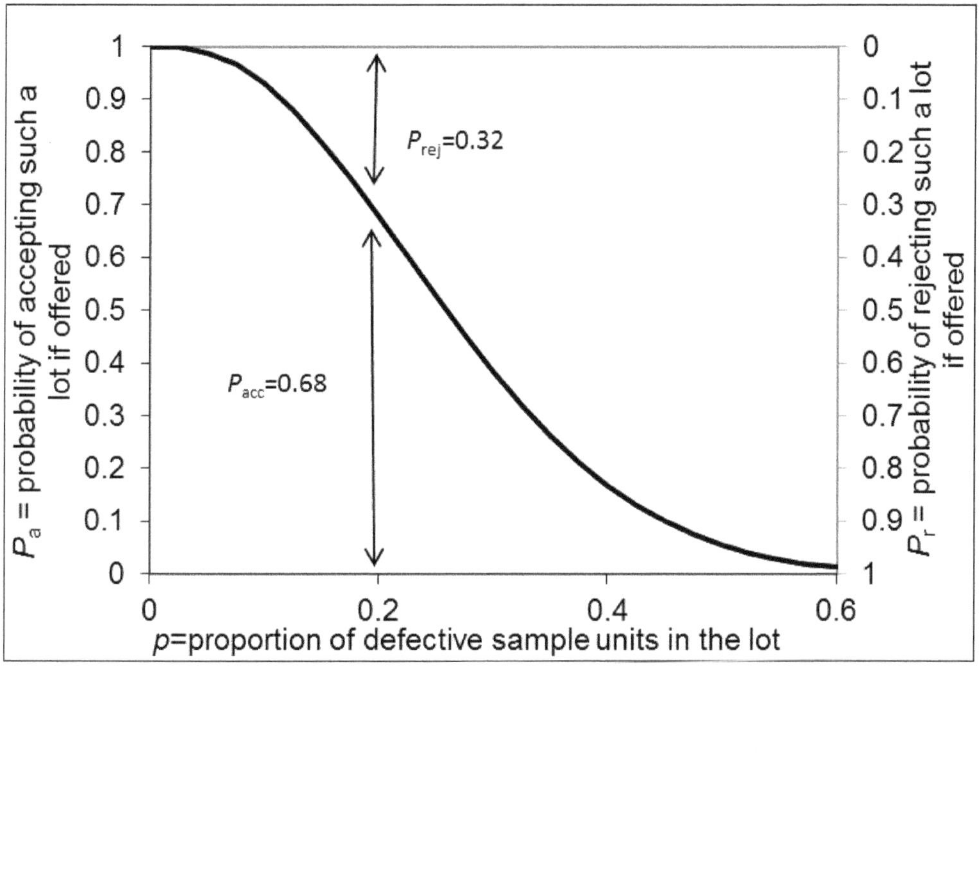

Three-class sampling plan – results can be satisfactory, acceptable or unsatisfactory

n = number of samples analysed;

m = maximum level of target microorganism(s) acceptable under conditions of good manufacturing practice;

M = level of target microorganism(s) or toxin(s) which, if exceeded, is considered unacceptable i.e. defective;

c = the maximum number of samples that can fall between m and M without the batch being considered unacceptable.

Levels falling between m and M are sometimes referred to as marginally acceptable and the stringency of a sampling plan depends on 'n' and 'c' e.g. for a given 'n', if 'c' is increased the plan becomes more lenient; if 'c' is decreased then the plan becomes more stringent (ICMSF, 2018). In industry, 3-class sampling plans are applied when using enumeration tests.

Sampling Plan Considerations

No sampling plan can ensure the absence of a particular microorganism. The choice of sampling plan should consider:

- The risk to public health associated with the hazard;
- The susceptibility of target consumer group(s);
- The heterogeneity of distribution of microorganisms, including the physical state of the food, e.g. solid or liquid;
- The randomness of sampling;
- The acceptable quality level;
- The likely prevalence of the hazard in materials tested and the desired statistical probability of accepting or rejecting a non-conforming lot;
- Cost-benefit of the plan and its purpose.

To decide whether the sampling plan should be two-class or three-class, there must be consideration of whether any positives can be permitted in the sample units. If the answer is no, then a 2-class plan should be used with c = 0. If the answer is yes, then a two- or three-class plan can apply and if the number of microorganisms in a unit-volume or mass can be measured, a three-class plan is recommended. Three-class plans may be more appropriate to accept several sample units with test results in the marginally acceptable interval and are more useful for trend analysis. They are also more appropriate when experience defines a different level above which counts indicate a potential risk for food safety.

Detection (presence/absence) Tests

These tests are used to detect a particular microorganism in a defined quantity of sample. Typically, these are used for detecting pathogens in foods. The quantity and number of samples are the two main considerations when determining the probability of detecting a target microorganism with a given frequency of contamination. This type of test is not usually used to quantify bacteria in a food however, data from tests for the detection (presence/absence) of specific microorganisms carried out in replicate may be used to ascertain the maximum probability of contamination of the food if all the results are negative or, the Most Probable Number (MPN) of target microorganisms present as cfu/g or /ml (see Section 4.3) if one or more result is positive.

If a high degree of confidence is required to detect very low numbers of positives (defects) in a batch, many samples must be taken (ICMSF, 2018). For practical and economic reasons, it is common practice to combine several samples to reduce the number of tests required. The only fully accepted method for this is to inoculate pre-enrichment broths and after incubation, to inoculate equal volumes from each pre-enrichment into enrichment broths. This allows lab personnel to go back to the original pre-enrichments in case a positive is detected. Further details on compositing are provided by Jarvis (2007).

Continuous sampling devices are useful for removing large numbers of small samples, particularly for powders and liquids, which are then mixed thoroughly before removal of test portions for examination.

Enumeration (counts) of Microorganisms

Enumeration procedures are normally carried out using a minimum test portion of 10g. A homogenate and further dilutions are prepared from the test portion and used to enumerate the level of microorganisms in the portion; counts are typically expressed in results as cfu/g. Estimates based on cfu/g assume that perfect mixing has occurred at all stages of sampling before any aliquot is removed for examination.

ICMSF three-class sampling plans are often used to describe specifications for microbiological counts. For example, where five samples of a ready-meal are taken and it is acceptable for the aerobic colony count to exceed 10^3 cfu/g on 2 occasions but not to exceed 10^4 cfu/g at all, the following applies:

$n = 5$, $m = 10^3$ cfu/g, $M = 10^4$ cfu/g, $c=2$

In many manufacturing situations the sample numbers taken and examined do not readily lend themselves to sample attributes plans; this may occur for instance, when a microbiological specification relates to one sample only ($n = 1$). In such cases, test results obtained from samples taken and examined from different batches could be used for trend analysis.

Censored Values

A censored value is one that is outside a measurable range. These are common in food microbiology. The commonest examples are 'less than' estimates e.g. $<10^2$ cfu/g, although 'greater than' examples ($>10^3$ cfu/g) do occur. It is common for industry to interpret 'less than' values as a measure of product acceptability and 'greater than' values as a measure of unacceptability.

For statistical analysis of results, where censored values are reported, extending the dilution ranges examined in enumeration or 'count' tests can further inform.

4. CONSIDERATIONS IN CHOOSING MICROBIOLOGICAL TEST METHODS

4.1. General

Microbiological examination of foods and food related samples is undertaken most often, in relation to aspects of verification of HACCP-based food manufacturing production and distribution systems. Although this has increasingly been developed, particularly within and for, larger food businesses over the past 20 years or so, European legislation in force from January 2006, now requires procedures based on HACCP principles for all food businesses (EC Regulations Nos. 853/2004 and 2073/2005).

The reliability of the results obtained from microbiological tests depends on the technical competence of laboratory staff using validated and verified methods working in a well-designed, constructed and maintained laboratory facility equipped with appropriate tools and materials. Since microorganisms were first cultured in laboratories in the late nineteenth century, traditional culture methods developed have involved the extensive use of chemically defined media and some containing organic substrates e.g. egg-yolk, meat particles or blood. Towards the end of the twentieth century, alternative methods have been developed that do not rely solely on traditional culture (McMeekin, 2003; Jordan & Lavender, 2012) and novel alternative methods are still being, and will continue to be, developed. Whichever methods are selected for use in a laboratory for the examination of foods and food related samples, they must be demonstrated as 'fit for the purpose intended' by those intending to use the method. This includes verifying the efficacy of the method compared to the laboratory's routine method, and preferably the current National or International Standard method (usually an ISO method or equivalent), for detecting the target microorganism in the product ranges examined routinely in the laboratory when the method is used by the regular laboratory staff. Such demonstration may also involve the use of samples inoculated with known levels of known microorganisms and samples received from an external quality assurance (EQA) test provider (Bell *et al.*, 2005).

Some assurance, to both the laboratory's staff and their customers, that appropriate standards of operation and practice are being maintained should be demonstrated by:

- Subscription to a laboratory accreditation scheme to provide an independent assessment, by audit, of the laboratory's quality management and operational systems;
- Subscription to EQA scheme(s) to provide an independent assessment of the laboratory's operation in practice including the laboratory's staff sample handling practices from receipt to reporting results alongside media preparation and routine quality control procedures as well as method efficacy, by means of a practical challenge and;
- The conduct of routine, internally developed audit programmes, involving all laboratory staff, particularly of practices and procedures affecting the quality of work carried out and also, internal quality control and assurance systems (IQA) for checking media quality and efficacy and full method efficacy using test samples inoculated with known levels of reference cultures (Bell *et al.*, 2005).

4.2. MAIN MICROBIOLOGICALLY DERIVED TOXINS ASSOCIATED WITH FOODS

The primary aim of this guidance document is to discuss and provide information and guidance concerning the microorganisms that food industry microbiologist and food regulatory authorities consider when developing food product safety/hygiene specifications and standards respectively.

It should be remembered however, that toxins may also need consideration in the context of a microbiological criterion. Some foods contain natural toxins e.g. phytohaemagglutinin in red kidney beans, tetrodotoxin in puffer fish, the risk from such toxins can be minimised through careful handling and processing of the foodstuff and these are not considered in this review. However, there are also toxins associated with food that are produced by a range of different microorganisms including bacteria, moulds and algae/dinoflagellates. Examples of the main types of such toxins and the microorganisms involved in their production are provided in Table 2.

Note: Routine testing laboratories do not generally examine foods for the presence of these types of toxins. Most of such testing is undertaken in laboratories with specific expertise of the test methods. Where a laboratory undertakes to examine foods for e.g. staphylococcal enterotoxins then, it is important to ensure that the method used will reliably detect the specific toxin of concern in the food being examined.

The potential for a food to be at risk from toxin-producing microorganism should always be considered during the hazard analysis undertaken at the product development stage. This is when raw material types, procurement and specifications and the specific processing conditions to be applied should be assessed for safety e.g. mycotoxins in cereals etc. Controls should always be in place to prevent growth of the microorganism to toxin forming levels as testing for the toxin should not be considered a satisfactory routine control.

Some additional information is given in Appendix I concerning toxins produced by specific foodborne microorganisms. The information in Table 2 and Appendix I provides examples only and is not intended to provide a comprehensive overview of microbial toxins. For further detailed information on such toxins please see Juneia & Sofos (2010).

The following sub-sections of this book concern microorganisms only and not their toxins.

Table 2: Some microbiologically derived toxins and their main food associations

Microorganism	Toxin type	Main food associations
Clostridium botulinum	Botulinum toxins Types A - G	Low acid canned foods containing e.g. vegetables, fish, meat that are under-processed. Minimally processed canned, bottled or preserved foods. Vegetable in oil products. Cook-chill and ready meals.
Bacillus cereus	Emetic	Cooked or fried rice, pasta and pastry products
	Diarrhoeal	Meat, soups and sauces
Coagulase-positive staphylococci	Staphylococcal enterotoxins Types A – E, G, H-R, U, V	Handled and temperature abused foods e.g. cooked sliced meats, dairy products, salads
Aspergillus flavus	Aflatoxin	Peanuts, treenuts, oilseeds and maize
Fusarium graminearum	Deoxynivalenol	Cereal crops
Penicillium expansum	Patulin	Apple juice
Spoilage microorganisms	Histamine	Scombroid fish e.g. mackerel, tuna
Algae/ Dinoflagellates	Paralytic shellfish poisoning (PSP) toxins e.g. saxitoxins; Diarrhetic shellfish poisoning (DSP) toxins e.g. okadaic acid; Amnesic shellfish poisoning (ASP) toxin (domoic acid); Azaspiracid shellfish poisoning (AZP)	Bivalve molluscs

4.3. CHOICE OF MICROBIOLOGICAL TEST

Test methods used in the microbiological examination of foods are designed either to detect very low numbers e.g. 1 viable cell per 25g, of a target microorganism or, enumerate levels where these are expected to be readily countable e.g. >10 cfu/g.

The microorganisms being detected or enumerated are effectively defined by the media and methods and techniques used. For example, an aerobic colony count (ACC), standard plate count (SPC) or total viable count (TVC) will only allow the determination of the number of microorganisms capable of forming visible colonies after a specified time on the growth medium used and under the specific temperature and atmosphere conditions used for incubation. Thus, an ACC at 30°C for 48 hours may give different information from an ACC at 22°C or 37°C for 48 hours.

Factors to be considered when choosing microbiological media and methods include:

- Proven capability of recovering the target microorganism(s) from the product being examined (through validation);
- How soon the result is required;
- Ability to achieve the sensitivity and specificity required;
- Cost-Benefit.

Information about relevant target microorganisms and the basis for their use in microbiological specifications can be found in Appendix I Specific microorganisms and the principles of test methods.

(a) Indicator and spoilage microorganisms

Examination of a product for the presence of 'indicator' (e.g. Enterobacteriaceae) or spoilage microorganisms can provide simple and reliable information about processing failure, post-processing contamination, contamination from the environment and the general level of hygiene. The methods used normally involve an estimation of the numbers of microorganisms in the food or environmental sample. These basic methods cannot replace examinations for detecting specific pathogens where suitable methods exist for detecting these and where such testing is appropriate. They do, however, usually provide information in a shorter time than that required for the detection, isolation and identification of specific microorganisms or pathogens, and the methods are often simpler and easier to perform.

(b) Bacterial pathogens

Most methods used for the examination of food samples for the presence of bacterial pathogens aim to detect very low numbers e.g. a single viable cell of the target microorganism, in a defined quantity of sample. Although detection tests are also referred to as presence/absence tests, absence of a pathogen from a food cannot be proven; it can only be demonstrated that a microorganism is 'not detected' by the test method used for the test sample type and quantity.

Classical microbiological methods used for the detection of bacterial pathogens are often slow and labour intensive. These methods often include a broth-based resuscitation step for sub-lethally injured microorganisms followed by selective broth enrichment and then isolation of the target microorganism using agar media that will encourage the growth of the target microorganism whilst suppressing growth of the background flora. This is then followed by individual biochemical and/or serological confirmatory tests on purified cultures of the suspect isolates. Because detection tests generally incorporate a resuscitation stage, they allow better recovery of stressed microorganisms than most direct agar plating methods.

4.4. TEST METHODS

It is important to ensure that the media and methods used will reliably detect the microorganism(s) of concern in the food being examined. It may be necessary to validate the efficacy of a method before it is used for the examination of test samples (see 4.1). When in routine use, quality assurance systems must be employed to verify continued test method and media efficacy i.e. through the use of appropriate EQA/proficiency testing schemes and the application of IQA systems.

(a) Competitive growth

The microflora present in foods can vary widely, for example between the freshly manufactured product and the same product at the end of shelf-life. Growth of specific target microorganisms may be suppressed (or they may be killed) by constituents of the food, its packaging and/or storage conditions or, they may be outgrown by other microorganisms present in the food. Carefully designed method validation studies may be necessary to demonstrate that the method selected for use will detect the target microorganism(s) in the presence of the competitive flora in the products to be examined.

(b) Comparability of results between different methods

A wide variety of media and methods are available for the growth of specific genera, and groups, of microorganisms e.g. indicator microorganisms. It is important to recognise results obtained from the use of different media and methods to detect or enumerate a particular microorganism(s) may not be comparable. This may not matter for the purposes of in-house trend analysis (if the same method is used consistently), but it is essential if the results are to be measured against, for example, industry or regulatory criteria (EC Regulation No. 2073/2005), using the reference method or a validated alternative. To this end, some uniformity of methods i.e. demonstrable comparability, and sampling procedures is desirable, but this should not prevent the development and application of the many new or alternative methods becoming available (Jordan & Lavender, 2012).

Results obtained from the use of different media and methods for the detection or enumeration of a specific microorganism(s) should not be assumed to be equivalent without evidence obtained from verification or validation studies.

(c) Sources of methods

Reference and industry test methods are available from several sources e.g. British Standards Institution, International Organisation for Standardisation (ISO, Geneva, Switzerland), International Dairy Federation (IDF, Brussels), International Commission on Microbiological Specifications for Foods (ICMSF), Public Health England (PHE, UK) - formerly Health Protection Agency

(HPA, UK), Association of Official Analytical Chemists (AOAC, USA), American Public Health Association (APHA, USA), Food and Drug Administration (FDA, USA), Campden BRI (UK).

Microbiological standards are also included in legislation (e.g. EC Regulation No. 2073/2005) and these also prescribe the methods to be used, or how to evaluate equivalent methods. In the case of any dispute concerning the microbiology safety of a food, national or international reference methods are used as the 'gold standard' method so the results will better withstand professional scrutiny e.g. in a court of law.

Appendix I Specific microorganisms and the principles of test methods describes the principles of classical test methods and discusses their relevance and limitations for some of the more commonly specified target microorganisms.

New, alternative test methods are becoming available for the detection of bacteria and their toxins and their performance is generally measured by comparison against results obtained by classical methods, especially published standard methods. For some microorganisms e.g. viruses and protozoa, methods for routine application to foods are more complex.

Methods currently available for the detection and quantification of viruses in foods are described in ISO 15216-1:2017 (ISO, 2017a) which specifies a method for the quantification of levels of hepatitis A virus (HAV) and norovirus genogroup I (GI) and II (GII) RNA, from test samples of foodstuffs (soft fruit, leaf, stem and bulb vegetables, bottled water) or food surfaces. ISO 15216-2:2013 (ISO, 2013) describes a method for the qualitative detection of HAV and NoV genogroups I (GI) and II (GII), from test samples of foodstuffs or food surfaces. Methods for detection of *Cryptosporidium* oocysts and *Giardia* cysts on berry fruits and leafy green vegetables are described in ISO 18744:2016 (ISO, 2016). The methods are based on fluorescence microscopy and are suitable for both qualitative and quantitative detection.

4.5. LIMIT OF DETECTION OF THE METHOD

The lower limit of detection of a test method is often referred to as its sensitivity. Plate count-based methods are generally used for enumerating levels of microorganisms where these are expected to be more than 10 cfu/g, and are more accurate when levels exceed 100 cfu/g. Most Probable Number (MPN) determinations are suitable for low counts e.g. <100 cfu/g, and are widely used for estimations of levels below 10 cfu/g. Detection (presence/absence) tests are used if information concerning the presence of a microorganism in a specified quantity of food such as 1g or 25g is required. The sensitivity of these tests is then defined by the quantity of food examined i.e. the greater the quantity of food examined, the greater is the test sensitivity.

4.6. Accuracy of the Result

Result accuracy depends on the method(s) used, media quality and technical competence of the laboratory staff as well as, the number, physiological state and distribution of the target microorganisms present in the test portion. ISO 7218 (BSI, 2015) provides tabulations of the statistical variability of estimated counts obtained from plating methods and MPN procedures. An understanding of the capability of the method and the error margins (uncertainty of measurement) of the results is essential for the sensible application of microbiological criteria and interpretation of results obtained. A count of 10^4 cfu/g appears very exact but could have been calculated from only ten colonies on a medium inoculated with 1ml from a 1 in 1000 dilution of the test portion. There are clearly defined rules relating to the number of colonies acceptable for counting under such circumstances (BSI, 2015), but these cannot always be followed. The lower accuracy of estimates made outside this acceptable range, e.g. very low numbers of colonies (typically lower than 20) needs to be recognised. The confidence limits of 'counts' ranging from 3 - 320 are illustrated in Table 3. Higher numbers of colonies counted are associated with greater precision of the final result.

ISO 7218 (ISO, 2017) also states that counts below 10 but greater than 3 should be calculated as and reported as an estimated count and that counts <4 should be reported as at the limit of enumeration. However, this level of accuracy should be placed within the context of the uncertainty of measurement of the test method which renders it usually insignificant when compared to the uncertainty from the sampling alone.

Table 3. Confidence limits associated with the number of colonies counted on a single plate

Colony count	95% confidence intervals for the count	
	Lower	Upper
3	<1	6
5	<1	9
10	4	16
12	5	19
15	7	23

MPN estimates are often used by food microbiologists but are not precise estimates of numbers of microorganisms. Nevertheless, MPN techniques are widely used (Mossel *et al.*, 1995) for samples in which levels of target microorganisms are expected to be low i.e. <10 cfu/g.

4.7. Recovery of target microorganism

Microorganisms may be damaged by the effects from a variety of physical processes and/or chemicals used during the manufacture of foods, including heating, freezing, drying, exposure to high osmotic pressure, low pH, disinfectants and preservative compounds. Although injured, cells may still have the potential to pose a hazard to consumer health.

Injured cells may recover poorly, or not at all, if exposed directly to selective media or elevated incubation temperatures without first being allowed to undergo cell repair using resuscitation techniques. The conditions and time for resuscitation required will depend on the target microorganism and the product sample characteristics/composition.

Plate count methods using a surface inoculation technique e.g. spread plate or spiral plate, may be preferable to methods using pour plate technique because of the possible injurious effect of the temperature of the molten agar used for the latter technique. In addition, some food-associated microorganism requires aerobic conditions for growth e.g. moulds and most *Pseudomonas* spp., and their growth may be impaired by the relatively anaerobic conditions obtained in pour plated media. These factors may result in a reduced count in comparison with agar surface methods of enumeration. The use of a liquid medium will normally provide better conditions for recovery than a similar solid medium.

There is evidence that some bacterial cells can become 'non-culturable' while remaining viable, that is, they cannot be grown under laboratory culture conditions in vitro, but are able to grow in vivo. This appears to occur in adverse (to growth) environmental conditions e.g. when the temperature decreases. Examples of genera containing species for which this has been demonstrated are *Salmonella, E. coli, Campylobacter* and *Vibrio*. Currently there are no routine laboratory methods for detecting microorganisms in this physiological state and any significance to public health has yet to be determined.

4.8. INHIBITION OF MICROORGANISMS GROWTH BY THE CONSTITUENTS OF THE FOOD TESTED

Growing microorganisms in laboratory culture is often affected by the chemical composition of the food under examination. If insufficiently diluted or neutralised, these chemicals may prevent growth of the target microorganisms during the test procedure.

Salt or sugar content, presence of chemical preservatives and the acidity of the product are examples of the chemical components of foods that may influence test results. Chocolate, many herbs and spices, garlic and onion all contain microbial inhibitory compounds such as essential oils and polyphenols in cocoa, and, many flavouring materials may contain high levels of salt and/or acids.

These inhibitory compounds may adversely affect the results of microbiological tests unless nullified during the test procedure. Method efficacy should be checked specifically, to ensure that target microorganisms present in such products will be detected, by examining test portions inoculated with known levels of representative strains of the target microorganisms thus helping to verify both media and method efficacy (Bell *et al.*, 2005).

4.9. QUALITY OF ANALYTICAL TEST RESULTS

For any test result to be considered valid, there must be confidence in the laboratory that undertakes that testing work. The testing laboratory itself should have appropriate quality systems in place that ensure that the work done is consistent and yields valid results that can be acted upon.

There are various parts of a quality system that all laboratories must have. These are:

- A suitable laboratory environment that minimises risks of environmental contamination, is easily cleaned and is separated from areas that handle food for consumption.
- Laboratory equipment that is well maintained, serviced and calibrated regularly. Records of servicing and calibration should be kept.
- Documented methods and laboratory procedures, that help to ensure standardisation and consistency of laboratory operations.
- Staff that are fully trained in all procedures and methods that the laboratory undertakes, with training records to confirm this. All staff should take part in regular competency testing to show their continued ability to perform and obtain the correct results. Records should be kept.
- All consumable materials (e.g. media, disposables etc.) should be quality control checked by the laboratory. Media required checking to prove they will allow correct target organisms to grow and prevent growth of non-target organisms.
- There should be procedures in place to minimise the risk of any laboratory cross contamination, such as the separation of activities by space or time.
- Methods that the laboratory uses must be valid for use. This usually means the use of recognised standard methods (e.g. ISO, CEN) or methods that have been fully validated against a standard method, to show equivalent results. Validation would usually be done against the

method validation standard ISO 16140 part 2, and be certified by organisations such as AFNOR, MicroVal or NordVal.

- All methods whether a standard, or a non-standard but validated method, must be verified by the laboratory. This requires the laboratory to do its own checks to ensure that it can use the method and obtain correct results in their hands with their samples (in the future a new ISO standard, ISO 16140 part 3 will inform laboratories how to do this).
- The laboratory should take part in regular external quality assurance (EQA) proficiency testing schemes. Such schemes will send the laboratory samples containing a microflora unknown to them. They undertake testing and return results to the provider. Laboratories should show acceptable results in these tests for all tests done.
- The laboratory should operate a full cleaning system, ensuring that all laboratory areas are cleaned and disinfected regularly. This should be verified by hygiene checks within laboratory areas. Again, records should be kept.
- The laboratory should operate a procedure for recording, reporting and correcting any non-conforming work (NCW). Any NCW noted by the laboratory must be investigated and a root cause found, then corrective actions put in place to prevent the issue re-occurring.

When using third party contract testing laboratories, it should be ensured that these are accredited to ISO 17025 (General requirements for the competence of testing and calibration laboratories). This accreditation will ensure that the points noted above have been met.

If a laboratory obtains a positive result for a pathogen, there should be enough checks done to give confidence that the result is correct and not an outcome of an error or cross contamination from another sample or a laboratory culture (e.g. a positive control strain). This may include checking media records, laboratory hygiene results, or, if Salmonella is involved, testing the isolated strain with antisera to the laboratories control strain (which should be an infrequently isolated serotype) in order eliminate control strain cross contamination as a reason.

Often laboratories will, or are requested to, retest samples that have been found to contain a pathogen. On many occasions, due to low level contamination of the sample, or lack of sample homogeneity, retests will give a negative result. In such circumstances the initial positive result cannot be discounted and must be considered to be true, unless a valid scientific/technical reason to discount this result can be established.

5. FOOD CATEGORIES

5.1. INTRODUCTION

In establishing microbiological criteria, it is important to consider the technologies applied in the preparation of the food. The following sections discuss some aspects of the technologies used and their effect on microorganisms.

5.2. FROZEN FOODS

Frozen foods are microbiologically stable i.e. there is no growth of microorganisms providing the temperature is below about -10°C. Most bacteria do not grow below 0°C, but some psychrotrophic microorganisms, particularly some fungi, will grow slowly at a few degrees below freezing. The shelf-life of frozen foods is determined by physical and chemical/enzymic changes e.g. dehydration and oxidation, and not by microbiological growth. Some bacteria and fungi die in a frozen food and others become sub-lethally damaged. However, most microorganisms are preserved by freezing; *Salmonella* spp. for example have been recovered from frozen foods after years of storage. Therefore, it must not be assumed that freezing can improve the microbiological safety or quality of a food. Freezing does not inactivate the toxins of microorganisms but in particular conditions, can reduce the viability and/or infectivity of some pathogenic protozoa and nematodes e.g. in raw meat and fish (ICMSF, 1980, Archer, 2004).

The application of microbiological criteria to frozen foods is dependent on the intended use of the food e.g. more stringent criteria may be applied to frozen foods which are ready-to-eat after being thawed.

Food should be handled and stored hygienically before freezing so that its quality does not deteriorate. Microbiological testing is occasionally used to monitor the efficacy of such handling processes. The rate of freezing and thawing can significantly affect the microbiology of a food and factors affecting the rate of freezing such as pack size and temperature gradients must be considered.

Frozen foods are often examined after controlled defrosting of the sample and microbiological tests used should incorporate a resuscitation procedure to maximise the recovery of any damaged cells.

5.3. Dried foods, low water activity foods and other ambient stable foods

Dried foods with an a_w of < 0.6 are microbiologically stable. The shelf-life of these foods is not limited by microbial spoilage provided they remain dry. A wide range of foods often termed 'intermediate moisture foods' or 'semi-moist' foods (jams, dried and fermented sausages, dried fruits) have a higher a_w (0.7 - 0.9). These foods are stored at ambient temperature and may be preserved by a combination of factors such as a_w, pH and preservatives; their shelf life may be limited by microbiological, chemical and/or physical spoilage.

Drying without heat (e.g. salting of fish) preserves many bacteria so the process cannot be relied upon to improve microbiological quality; in fact, the net effect can be to increase the concentration of microorganisms. Numbers of microorganisms generally decline slowly in dried foods (over a period of months or years) although *Salmonella* spp. have been shown to survive for years in chocolate (ICMSF, 2005) and powdered milk and spore-forming microorganisms will also remain unaffected by these dry environments. Due to its fastidiousness and oxygen sensitivity, *Campylobacter jejuni* is not normally expected to survive well in oxygen-rich, atmospheric conditions. However, recent studies (e.g. Oh *et al*., 2015) investigating tolerance to aerobic conditions indicate that some strains of *C. jejuni* can become hyper aerotolerant, surviving much longer than aero-sensitive strains and there is some suggestion that these may be more virulent than aero-sensitive strains (Oh *et al*., 2017).

Microbiological criteria for dried foods are varied and take account of the intended use of the food (ready-to-eat or not), the intended consumer (baby foods or other vulnerable groups) and the type of food (heat processed, fermented).

Prior to the microbiological examination of these product types, the inhibitory factors in many products (see Section 4.7) must be diluted out or neutralised prior to microbiological examination (Roberts & Greenwood, 2003).

Pathogens present in dried foods, such as *Salmonella* in herbs and spices, may be stressed and a resuscitation stage must be considered in detection procedures. In some situations, conditions may change in the food matrix e.g. moisture ingress into a dry food, and it may be possible that such changes may result in conditions supporting growth. It may therefore also be important to measure the water activities of food samples to confirm the conditions in the food.

5.4. FRESH & CHILLED FOODS

Most fresh foods are not microbiologically stable, and some commodities are highly perishable. The associated microflora and consequent shelf-life will be affected by the quality and hygienic conditions of animal husbandry and slaughtering practices, harvesting, processing and storage. The application of rigorous hygiene measures at all stages will help to prevent product contamination and achieve a realistic shelf-life. Extension of shelf-life of many fresh foods can be achieved by chilling, and the type of packaging used may also have an effect.

During chilled storage, the microbiological profile will change from predominantly mesophilic to psychrotrophic microorganisms, and spoilage will be due to the nature and activity of these psychrotrophs. In contrast, the microbiological quality of foods such as nuts and citrus fruits that have a natural protective outer casing depends on maintaining the integrity of that outer barrier and the control of moisture during storage. The contents are comparatively microbiologically stable unless penetration of the outer casing occurs.

The occasional presence of pathogens can be expected in most raw foods. When selecting relevant microbiological criteria, the type of food, the required shelf-life, storage conditions, intended use and destination of the commodities must all be considered.

Aerobic colony counts may be used to indicate the quality and the probable shelf-life of many raw fresh foods; whilst normally this is performed at 30°C to measure the mesophilic population, a temperature of 20-25°C may be preferable for chilled foods to estimate the levels of psychrotrophic microorganisms present and hence the potential spoilage microorganisms.

Yeast or mould counts are more relevant indicators of shelf-life than bacteria for commodities with a low pH such as fruits and fruit juices or commodities with reduced a_w. Indicator microorganisms may be used to monitor the hygiene of processing and may also give information about the temperature conditions of storage. In addition, *E. coli*, Enterobacteriaceae, or coliform levels may be useful in monitoring the quality of irrigation water used for fruit and vegetables, to monitor slaughterhouse hygiene and the conditions for production and marketing of live bivalve molluscs.

5.5. HEAT-TREATED FOODS

Heat-treated food may be ready-to-eat either hot or cold, may require reheating or further cooking before consumption, or may undergo further cooking as an ingredient in a composite food product. The degree of heat treatment it has received will affect its microbiological profile and stability. Some forms of heat treatment such as blanching or scalding are not designed to cook the food, but to inactivate enzymes or assist in skin or shell removal. This type of mild heat treatment and others such as pasteurisation and sous-vide processes will kill most vegetative bacteria, leaving only heat resistant species/strains and spores, or their heat-resistant toxins. Pasteurisation or an equivalent process also kills most of common foodborne yeasts and moulds but does not destroy heat-resistant moulds and mycotoxins. Testing heat-treated foods for indicators such as Enterobacteriaceae or coliforms can prove useful for verifying the effectiveness of certain processes or provide evidence of post process contamination (see below).

Growth of any surviving microorganisms in many heat processed foods is minimised by chilled storage, acidity, reduced a_w, chemical preservation, or some other controlling factor.

Long-term microbiological stability without the use of refrigeration for non-dried foods is obtained by prolonged heating in hermetically sealed containers such as cans, jars or foil pouches; the heat process will need to be more severe for low acid foods (Hersom & Hulland, 1980). Low acid foods (pH greater than 4.5) that do not contain other combinations that are able to control *C. botulinum* (such as those in processed cheeses for example) must receive a minimal thermal process of F_0 3 (121.1°C for 3 minutes) known as a 'botulinum cook' or an equivalent heat process. Higher thermal processes are often required to minimise spoilage due to thermophilic spore-forming bacteria in canned foods destined for or manufactured in countries with warmer climates, and it may be necessary to apply processes with $F_0 > 10$.

Note: A F_0 value can be defined as: "A measure of the amount of lethal heat which results from a specified thermal process in the slowest heating part of the can. The number is the lethal effect equivalent to the number of minutes at 121.1°C (250F) when assuming instantaneous heating and cooling and a z value of 10°C degrees (18 F degrees)" (UK Department of Health 1994).

A more severe heat treatment such as the Ultra High Temperature (UHT) process used for certain liquids results in a shelf-stable product at room temperature provided that aseptic filling occurs after heat treatment or that the liquid is heated in the final packaging.

Oven-baking of biscuits and certain cakes may also result in stable products at ambient temperature due to the low a_w, as long as contact with moisture and subsequent surface mould contamination can be prevented. Cooking of meats and meat products may allow the survival of bacterial spores, and so chilled storage will be necessary after cooking.

The bactericidal effect of heat will depend on the nature of the food, for example the fat or sugar content, moisture content or acidity, as well as the temperature and exposure time employed. The parameters of the heat treatment must take account of the characteristics of the food. Controlling and monitoring the temperature and duration of heat treatment in all these processes is of prime importance; the integrity of the containers is also essential to prevent ingress of contamination after heat treatment. The main purpose for microbiological testing of heat processed foods is to assess post-processing contamination. Ingredients added after heat processing may affect the microbiological status of the product and hence the criteria that are applicable.

The nature and extent of contamination introduced by personnel and equipment after heat treatment will affect the resulting shelf-life and safety of the food. Microbiological criteria should reflect the intended use of the commodity; for example, the specification for a cooked ingredient of a composite product such as pizza which receives a further cooking process might be less rigorous than if no further heat process is to be applied. Consumer behaviour, cooking instructions, the type of packaging and expected shelf-life should also be taken into consideration.

For example, Codex (Codex, 2007) advises that testing against microbiological criteria for *L. monocytogenes* may not be useful for:

- Products that receive a listericidal treatment after being sealed in final packaging that ensures prevention of recontamination until opened by the consumer or otherwise compromised;
- Foods that are aseptically processed and packaged, and;
- Products that contain a listericidal component that ensures rapid inactivation of the pathogen if re-contaminated (e.g. products that contain > 5 % ethanol).

Listeria monocytogenes should be absent from products which allow its growth within at least the expected shelf life of the product, under reasonably foreseeable conditions of distribution, storage and use during the shelf-life. However, a low level of *Listeria monocytogenes* might be tolerated in a chilled food likely to be eaten within a few days. EC Regulation (EC No 2073/2005) requires that numbers of *L. monocytogenes* do not exceed specified limits, i.e. 10^2 cfu/g in ready-to-eat-foods, absence in 25g in those foods intended for infants and for special medical purposes. Food business operators need to be able to provide evidence to demonstrate that food products will not contain levels >10^2 cfu/g of *Listeria monocytogenes*.

5.6. Fermented Foods

The term fermentation describes the process of intended chemical change in foods catalysed by microbial enzymes. Fermented foods may be produced with the use of starter microorganisms or by the action of natural fermentative microorganisms present in the raw materials or production environment.

Many microorganisms are used in fermentation processes but those most widely used in the food industry include lactic acid bacteria and yeasts.

The metabolic products of fermentation processes often allow the extension of shelf-life and enhance the safety of food products because of their inhibitory effect on pathogenic and spoilage microorganisms. Lactic acid bacteria used for yogurt production produce organic acids which decrease the pH and act to restrict the growth of both spoilage microorganisms and pathogens. The original preservation systems developed for fermented raw meat products were based on low pH, reducing a_w (reduced water content and/or added salt), addition of nitrite and subsequent storage at a minimum temperature/time (often referred to as the 'quarantine period') to eliminate infectious pathogens and prevent growth of toxigenic microorganisms and spoilage organisms. These factors, together with raw material quality (to ensure infectious vegetative pathogens are at low levels or absent) are still used as the basis for assuring safety of several fermented raw meat products.

There are also products where the manufacturing process involves a terminal pasteurisation step to destroy vegetative infectious bacteria (e.g. for products that are chilled or consumed directly after manufacture i.e. no quarantine period). It is critical that manufacturers of these products where no terminal pasteurisation step is applied review relevant significant hazards on a regular basis to ensure that existing controls remain effective. For example, with the emergence of STEC in raw beef, several outbreaks occurred in the 1990s demonstrating that existing controls were not effective. With emergence of hepatitis E virus in pork meat, this also requires careful review by manufacturers of these products to assure their safety.

It is also critical that manufacturers monitor key controlling factors during validation of the processes, to ensure that appropriate limits are set and followed during routine manufacturing.

Some starter culture microorganisms also produce bacteriocins which have an inhibitory effect on other bacteria. Yeasts used for wine and beer production produce organic acids and ethanol, which create an extremely harsh environment where few microorganisms can grow.

Starter cultures may be susceptible to infection by specific bacteriophage (viruses) that can change the functionality of the culture and impact quality and food safety attributes on the finished product. Currently, there are no commercial starter cultures available which are completely resistant to all bacteriophage. Even if a culture appears phage insensitive it is usually possible to detect bacteriophage after a period of time due to the rapid evolution of the phage. The testing of potential production strains with a representative phage collection gives an indication of the behaviour of the strains. Strains which survive several cycles with a great number of purified phages are of good quality. Furthermore, pre-testing the culture at the process stage where the culture is to be introduced can also be used to predict how the starter culture will behave.

Fermented foods can also develop an accumulation of biogenic amines, such as histamine, from the decarboxylation of amino acids. Such a reaction is from the growth of bacteria which contain such decarboxylase enzymes. Whilst some low levels of amines can be tolerated, high levels can cause food poisoning. For this reason, starter cultures are selected to avoid those with the ability to generate such amines. However, the presence of natural organisms causing this problem cannot always be avoided such as in the use of raw milk for fermented dairy products. Acceptable limits are difficult to set but where applicable these are provided within this publication.

The raw materials used as substrates for fermentation reactions e.g. milk, wort and meat are equally capable of supporting the growth of many pathogenic and spoilage microorganisms, if present. The safety and stability of the fermented product is therefore dependent on the use of raw materials of good microbiological quality as well as the vigorous growth of the fermentation microorganism(s) to produce inhibitory compounds and the rate at which these compounds are produced. Strict process control is essential to ensure that the desired fermentation profiles are achieved. Slow growth, or no growth, of the fermentation microorganisms may allow pathogenic or spoilage microorganisms to predominate resulting in unsafe products or ones that do not meet the required organoleptic qualities.

Fermentations are often monitored using simple indicators of microbial growth such as decreases in pH or sugar concentrations or increases in cell numbers (turbidity, direct microscopic examination), acidity or alcohol concentrations. Fermentation alone can lead to inhibition of microbial growth and, in some cases, to microbial death although this will be dependent on both the type and concentration of metabolic products produced by the fermenting microorganisms and on other conditions, such as storage temperature. Storage at refrigeration temperatures can reduce the lethal effects of antimicrobial conditions and will often promote survival of microorganisms compared to storage at higher temperatures.

The application of microbiological criteria to fermented foods is dictated by the nature of the product and its intended use. For example, fermented products that are matured, like hard cheeses, do not afford the same conditions for survival compared to less mature/aged products such as soft cheeses made from unpasteurised milk. Sampling for particular pathogens e.g. *Salmonella* spp., *L. monocytogenes*, coagulase-positive staphylococci and STECs such as *E. coli* O157 needs to be considered for these products (see Section 6).

Criteria are frequently applied to a product at stages during the manufacturing process e.g. coagulase positive staphylococci in the production of salami, cheese and to the final product. Monitoring should be done during the process until the desired pH has been reached that will inhibit growth of the organism. A level found at $\geq 10^5$ cfu/g should lead to further testing for staphylococcal enterotoxins.

Levels of significance of specific microorganism groups in/on a particular food must be understood e.g. psychrotrophic moulds such as some *Penicillium* spp. can be commercially significant for block matured, vacuum-packed cheese if bags are punctured and the microorganism grows to spoil the product. In this circumstance, a significant contamination rate could be as low as one spore on the entire surface of a block of the cheese and a specification with criteria for moulds of <10 cfu per gram would be meaningless. Control requires ensuring the vacuum remains intact and the storage temperature is maintained.

In some cases, the acidity or alcohol content in fermented foods may need to be neutralised or diluted out prior to enumeration or detection of microorganisms as they may interfere with the indicator dyes used in some agar test methods e.g. Violet Red Bile Glucose Agar used for the enumeration of Enterobacteriaceae. In addition, test procedures for the detection of pathogens should incorporate a resuscitation stage as the effect of pH, acidity or alcohol content in the food may stress the microorganisms.

5.7. PACKAGING

Different packaging technologies are used for a various type of foods and for many, they are of paramount importance in maintaining the integrity of products and preventing recontamination.

Packaging materials vary and include metal, glass, plastic, card and paper, or combinations of these. Different packaging types vary in their physical and gas-barrier properties and in conjunction with the processing and storage conditions, will select for particular groups of microorganisms. However, having 'air-tight' packaging does not necessarily mean that microorganisms will not grow in products as conditions may develop that, for example, favour the germination of the spores of anaerobes and the subsequent growth of these microorganisms.

Some products are heat processed and then hot-filled into packs or cold-filled by aseptic or clean-filling methods while others are heat processed 'in pack'. With these products, validating that the process is under control is only possible using large numbers of samples (see below) and this is required when commissioning new equipment or a modified line, or following a breakdown in manufacturing. In addition, the seal integrity of the pack should be monitored, together with the contents of the pack. Spoilage or 'infection' rates will vary according to the process/packaging used and microbiological criteria applied need to take account of these, together with other factors, such as product formulation.

Due to the low residual numbers of viable microorganisms/spores expected immediately following heat processing, sampling plans for commercially sterile products or products that have undergone high heat processes will often require product packs to be pre-incubated at the optimum growth temperature of the target microorganism(s), for a specified period prior to microbiological examination.

In addition, pack incubation tests involving large numbers of packs may be carried out e.g. for aseptically-packed products and canned products, to provide positive assurance that product processing and packaging are under control. However, these incubation tests have limitations and must never be used as the sole or predominant criterion to assess safety. On-line process controls and records e.g. thermal process records remain the most important means of providing positive assurance of product safety.

6. MICROBIOLOGICAL LIMITS

6.1. INTRODUCTION

There have been a number of publications over the years referring to microbiological criteria for foods including Murray (1969), National Research Council - United States of America (1985), International Commission on Microbiological Specifications for Foods (1986), Shapton & Shapton (1991), Codex (2013a) but few of these incorporate modern food industry's approach to the development and application of microbiological criteria to foods. A guide setting out the principles and process for establishing and using microbiological criteria has been published by Campden BRI Guideline 52 (Campden BRI, 2017) and explanations as to the criteria to use when setting microbiological criteria can be found in the ICMSF book 8 (ICMSF, 2018).

Indiscriminate application of microbiological testing and criteria is not advocated and should be avoided. If, after due practical consideration of the issues raised and discussed in previous chapters it is considered useful to use microbiological tests and criteria for clearly specified purposes then the following tables of microbiological information may help as a further guide to the use of relevant criteria.

The figures given in the following tables do not represent microbiological criteria in themselves but should be used as the basis for the development of criteria by the food business operator. Use of internationally recognised or equivalent validated industry methods is assumed.

For easy reference, products are listed in the tables by type and technology, and then broadly split into groups based on their use e.g. to be cooked, ready-to-eat. A comprehensive list of foods with associated combinations of technology, packaging and use is not the aim of this document.

Guidance on limits and some key points concerning pathogens and indicator microorganisms relevant to each food grouping is given. Important toxins are also noted for some groups.

Some of the levels given in the Tables are based on existing published criteria (see Health Protection Agency, 2009). Others are derived from industry specifications, practice and experience. The limits given in these tables are considered by the authors to be practical, realistic and relevant to the use of the food and take account of the potential hazards associated with the food, processing practices and risk to the consumer.

It may not be necessary to apply all the tests listed for a product category and on some occasions, it may be relevant to test for additional microorganisms and/or toxins.

Although it may be desirable to minimise the level of specific microorganisms, criteria for microorganisms in foods that require cooking may be less stringent than those for ready-to-eat foods. In particular, criteria for *Escherichia coli* (not STEC) and coagulase positive staphylococci are often over-specified and seldom justified in terms of food safety or food spoilage. High levels ($>10^4$ cfu/g – Health Protection Agency, 2000) of coagulase positive staphylococci may be relevant due to their ability to produce heat resistant enterotoxins although low levels of the microorganism are usually of little significance in foods prior to cooking.

The Guidelines for Assessing the Microbiological Safety of Ready-to-Eat Foods Placed on the Market by Public Health England (HPA, 2009) gives reference to the safe/satisfactory level of *Bacillus* spp. in product before they become a concern of food safety risk for human health. As a general rule of thumb levels below 10^3 cfu/g is considered to be safe. Levels above 10^5 cfu/g are usually considered to be potentially injurious to human health and unfit for consumption. Although not all *Bacillus* spp. are hazardous, the most important recognised as such is *Bacillus cereus.*

Any raw food may be contaminated with pathogens such as *Salmonella* spp., *Campylobacter* spp., *L. monocytogenes* and STEC. Criteria for the absence of pathogens in raw foods intended to be eaten without cooking are of limited benefit and do not guarantee safety of these products. Absence of pathogens in products which are eaten raw is desirable but cannot be guaranteed (Huss, 2003). Food poisoning is also associated with raw products e.g. raw milk; therefore, the consumption of raw milk or raw (or undercooked) meat carries with it a significant risk, particularly for people with a weaker immune system such as infants, the elderly, pregnant women or those with underlying medical conditions. The application of HACCP-based systems and GMP at the farm and during harvesting, processing, storage and transportation is necessary to reduce (but will not eliminate) microorganisms in such foods. Likewise, microbiological monitoring for trends can be useful, but no amount of further testing can assure absence. Clear advice to consumers about the risks from these foods will also help to minimise the risk of foodborne illness.

All criteria should include a reference to the actions to be taken in the event of deviations. Such actions need to be pre-defined and as specific as possible to drive timely and appropriate interventions. The following tables do not include such actions as they are risk-based according to the specific application of the criteria, which is unique to the product, process and environment, apart from safety criteria, where the appropriate action is usually defined as those steps necessary to prevent harm to the consumer.

When defining actions to be taken it is important to realize the difference between process indicators (e.g. process hygiene criteria) and safety criteria. Actions taken upon failure to meet process hygiene criteria should not be to automatically reject product but should be to investigate further as such criteria are only designed to indicate a potential toward loss of control and do not in themselves indicate a food safety concern. Further actions may however be required upon investigating the root cause of the failure to meet the criterion.

On the other hand, failure to meet a food safety criterion usually indicates a potential for that foodstuff to cause harm and appropriate and timely actions are required to minimise such impact and regain control.

All actions must consider the past, present and future to anticipate the impact on previous production, continuing manufacturing and also actions required to minimise the risk of recurrence. Any and all actions must be fully documented and reviewed to ensure effective close out.

Criteria should also include a reference to the frequency of testing, where testing is the appropriate means to provide assurance against the criterion. Such testing frequencies are defined according to risk of failure and impact to the consumer.

Note: A criterion does not necessarily imply the need to test against it. Testing is often the weakest form of assurance due to sampling error associated with the variable and non-homogeneous distribution of contaminants. Failure to meet a criterion informs a great deal whereas meeting a criterion via testing informs very little about the microbial status of a foodstuff.

Where there is a known history of microbiological problems with a product type or a lack of knowledge of the microbiological hazards to support risk assessment, such as a new process or product type, a higher level of testing for appropriate pathogens or tests for other microorganisms may be necessary, but such testing can never provide adequate assurance by itself of the food safety status of a product.

Typically finished products' testing is performed monthly, with sampling taken throughout the month to cover all variables of process and/or product type. However, there is no set definition of a suitable frequency, often being set between manufacturer and customer. Where an excessive level of frequency is in place this often demonstrates a lack of understanding of the potential microbiological hazards or lack of confidence in the manufacturing process and an over estimation of how much assurance such testing can then provide.

Much greater assurance can be provided by direct monitors of specific controls applied during the process. Validation and verification testing of control measures is covered in more depth in ICMSF book 8 (ICMSF, 2011) and other guidance documents such as Codex (2013b), Brackett *et al.* (2014), and Campden BRI Guideline on validation and optimisation of thermal processing systems (Campden BRI 2015).

6.2. USE OF TABLES

For practical reasons the following tables represent broad product categories and inevitably some product groups may not conform to the limits given for indicator microorganisms. For some multi-component products e.g., sandwiches containing cooked meat and salad ingredients, the information in more than one table will need to be considered and the impact of interaction between components may require additional consideration.

The limits for food safety criteria given are based on the known levels causing harm and as such are well defined and supportable from epidemiological evidence. The limits for process hygiene criteria are not well defined as they may well be specific to not only the food category but also the method of manufacture, intended shelf life etc.

As such the values for the food safety criteria can be relied upon to define the point of acceptability and unacceptability whereas the limits given for process hygiene criteria are for guidance only and true values should be defined by the manufacturer to indicate when their process is in control and when there is a trend towards loss of control.

Whenever testing demonstrates a misalignment to the specification applied, it is the responsibility of the manufacturer to take the appropriate action. Such actions should, as far as possible, be pre-determined and documented to direct consistent, appropriate and timely actions to out of specification events.

As stated, for safety criteria the action must be to minimise the impact on the consumer by whatever mechanism is most appropriate, which ultimately is to reduce the amount of affected material available on the market and to communicate the concern effectively to all stakeholders.

In the event of failure to meet hygiene criteria, the action must be to investigate the cause, assess the impact and take all necessary actions to recover control.

The figures given in the tables are under two headings: **GMP (good manufacturing practice) and Maximum.**

- **GMP values** are those expected immediately following production of the food under good manufacturing conditions employing a HACCP-based system of controls, i.e. at the end of manufacture. Occasional samples may exceed these limits and such results should not be used as rejection criteria without consideration of all the factors important in an entire sampling plan (Chapters 1 and 3).

- **Maximum values** are those regarded as the maximum acceptable at any point in the shelf-life of the product, including at end of shelf life. Except for those pathogens when only absence is permitted, higher levels of certain pathogens may indicate a possible health risk whilst higher levels of indicators may signify failure of the process or hygiene procedures.

Note: The GMP values and Maximum values denoted here should not be confused with the 'm' and 'M' values that are used in sampling plans.

In many products, microorganisms may continue to grow throughout shelf-life. If 'Maximum' levels are approached early in the intended life of the product, it is likely that this will exceed the maximum levels within its shelf-life. This must be considered when using these tables.

The levels of any specific indicator microorganism present may vary according to food material and processing conditions. Maximum levels given in the tables indicating potential for process failure may therefore differ to reflect these variations. For example, products which have been processed to achieve a 'botulinum cook' are expected to remain safe throughout their shelf-life; low 'Maximum' criteria are therefore usually applicable to these products.

The values are intended to be applied to a representative sample of the food, which in most cases would be not less than 10g (25g for detection (presence/absence) tests for pathogens) and would include surface material, core material and a selection of the components in the case of a multi-component product. Ingredients added after the processing stage indicated in the tables are not considered and must be taken account of separately. Sampling plans and methods are not indicated in these tables and should be developed and used in line with the considerations discussed in Chapters 3 and 4.

Microbiological criteria should only be applied to foods after consideration of all the matters detailed in this document and, when applying microbiological tests and criteria to foods, all elements of the microbiological criterion including the sampling plan must be clearly specified. The effect of the time of sampling in relation to the assigned shelf-life of a product, its proposed end use etc. must also be recognised.

The European Union and some individual nations currently have legislation containing microbiological criteria applicable to some foodstuffs at defined points in the production process. Products identified in current EU legislation are listed in section 7. It is essential for food businesses to ensure that products conform to relevant current legislation of the countries in which their products are made and, marketed/sold.

Concerning mycotoxins, these are not included in the following tables and further information on limits for these hazards can be found in Regulation (EC) No. 1881/2006 (EC, 2006).

6.3. INDEX FOR FOOD TABLES

FOOD	TABLE
Crudités	F
Crumpets	M
Cured meats	B, K & R
Dairy desserts	G
Dairy ice cream - pasteurised	G
Dairy products - pasteurised	G
Dairy products - UHT	I
Dessert mixes	O
Desserts	K
Doughs	D
Drinks - soft	T
Drinks – fruit-based	E
Drinks – low/no-alcohol beers	T
Dry mixes	M & O
Dry mixes - for baby foods	P
Egg - liquid	J
Egg - dried	O
Fat spread – non-dairy	S

FOOD	TABLE
Fermented products - pasteurised	G
Fermented products - non-pasteurised	H
Fish - raw	C
Fish - cold smoked	C
Fish - hot smoked	K
Fish - pickled	C
Fish - marinated	C
Fish - dried	M
Fish - cooked	K
Fish - canned	Q
Flans- savoury	K
Flavourings	N
Flour	M
Fresh herbs	F
Freshly squeezed juices	E
Fruit - dried	M & N
Fruit - raw	E & N
Fruit - pasteurised	E

FOOD	TABLE
Fruit - preserved	R
Fruit cocktail	E
Fruit drink - heat processed	E & T
Fruit drink - preserved	T
Fruit juice - raw	E
Fruit juice - pasteurised	E
Fruit juice - freshly squeezed	E
Fruit squashes	T
Gravadlax	C
Gravy mixes	O
Ham - cooked	K
Ham - raw	B
Ham - dry cured	R
Herbs - dried	M & N
Herbs - fresh	F
Herbs - in oil	R
Ice cream - pasteurised	G
Infant formulas	P

FOOD	TABLE
Intermediate moisture foods (IMF)	R
Jams	R
Lard	S
Lemonade	T
Margarine	S
Meat - raw (joints, mince, diced)	B
Meat - cooked	K
Meat - canned sterilised	Q
Meat - cured	B & K
Meat - fermented	K
Meat - dried	N
Meat - dry cured	K & R
Meat - marinated	B
Meat - smoked	K
Milk - non- pasteurised	H
Milk - pasteurised	G
Milk - long life	I
Milk powders	O

FOOD	TABLE
Milk powders - for babies	P
Milk powders – non-pasteurised	H
Milk powders - pasteurised	G
Milk shake - UHT	I
Muesli	N
Muffins	L
Mussels	C & K
Noodles	D & M
Nut butter	S
Nut oil	S
Nuts	N
Offal	B
Oils	S
Oysters	C
Part-baked product	D
Pasta - fresh filled	J
Pasties	K
Pasteurised milk	G

FOOD	TABLE
Pastry	D
Pickles	I
Pies (meat, fish and or vegetable)	J & K
Pizzas	J & K
Pork	B & K
'Pot' snacks	O
Pouched foods	Q
Poultry - raw (whole, portions, minced)	A
Poultry - cooked	K
Poultry - canned	Q
Poultry - marinated	A
Poultry - smoked	K
Powdered desserts	O
Prawns - raw	C
Prawns - cooked	K
Preserved foods	R
Processed cheese	G
Pulses - dried	M

FOOD	TABLE
Pulses - cooked	K
Quiches	K
Recipes dishes/ready meals - raw	K
Recipes dishes/ready meals - part cooked	J
Recipes dishes/ready meals- heat processed	K & Q
Relishes	R
Rice	M
Rice products	K
Roll mop herrings	C
Salad vegetables – raw/prepared	F
Salami	J & R
Sandwiches & sandwich fillings	F & K
Sauces	K & R
Sauce mixes	O
Sausages	B
Scombroid fish	C & K
Shellfish	C & K
Smoked salmon	C

FOOD	TABLE
Snacks	N
Soft drinks	T
Soup - mixes	O
Soup - canned/pouch	Q
Sous vide	K
Spices dried	M, N & O
UHT dairy products	I
Vegetables - heat treated	K
Vegetables - raw/prepared	F
Vegetables- dried	M
Vegetables - preserved	R
Vegetables - frozen	F
Vegetables - stir-fry	F
Vegetables - canned	Q
Virgin olive oil	S
Wine	T
Yoghurt - pasteurised	G

6.4. TABLES

The microbiological criteria in these tables should only be applied to foods after consideration of all the matters detailed in this document and, when applying microbiological tests and criteria to foods, all elements of the microbiological criterion including the sampling plan must be clearly specified.

All numerical counts/levels given in the following tables represent colony forming units per gram or per ml unless otherwise specified.

ND = not detected

$10^n = 1 \times 10^n$ where n = 2, 3, 4, 5, 6, 7, 8 etc.

GMP values are those expected immediately following production of the food under good manufacturing conditions.

Maximum values are those regarded as the maximum acceptable at any point in the shelf-life of the product (see 6.1 and 6.2).

Note: The GMP values and Maximum values denoted here should not be confused with the 'm' and 'M' values that are used in sampling plans.

REMINDER: Indiscriminate application of microbiological testing and criteria is not advocated and should be avoided. It may not be necessary to apply all the tests listed for a product category and on some occasions, it may be relevant to test for additional microorganisms and/or toxins.

A: RAW POULTRY

Product examples: Whole birds, portions, minced/reformed poultry meat, marinated products .

Storage: Frozen or chilled
Use: To be cooked

Pathogens

Salmonella spp., *Campylobacter* spp. and other pathogens may be present; monitoring maybe necessary to meet legislative requirements

Product	Microorganism	GMP (end of manufacture)	Maximum (end of shelf life)
Minced meat, preparations and products	*Salmonella* spp.	ND in 25g	ND in 25g
	Salmonella Typhimurium	ND in 25g	ND in 25g
Fresh poultry meat	*Salmonella* Enteritidis	ND in 25g	ND in 25g
	Campylobacter spp.	10^3	10^3

Product	Microorganism	Process Hygiene Indicator
Poultry carcasses	*Salmonella* spp.	ND in 25 g (pooled neck skin)

Indicators & spoilage microorganisms

ACC can be used to indicate quality. Generally, higher counts are found in minced/reformed poultry meats than on whole birds or portions. Pseudomonads are also frequently used to indicate quality.

Microorganism	GMP (end of manufacture)	Maximum (end of shelf life)
ACC	$<10^5$	10^7
Pseudomonas spp.	$<10^5$	10^7
Yeasts (marinated products)	$<10^4$	10^6

B: RAW MEAT

Product examples: Joints, mince, diced meats, offal, burgers, sausages, bacon, marinated products, cured products (not ready-to-eat).

Storage: Frozen or chilled
Use: To be cooked

Pathogens

Salmonella spp. *Campylobacter* spp., STEC and parasites and other infectious agents may be present; monitoring the incidence of some bacterial pathogens may be necessary to meet legislative requirements.

Microorganism	GMP (end of manufacture)	Maximum (end of shelf life)
Bacterial pathogens e.g. *Salmonella* spp., STEC	Monitoring incidence may be useful for trend analysis	

Indicators & spoilage microorganisms

ACC can be used to indicate quality. Counts are generally higher in minced meats than on whole cuts. *E. coli* may be useful as an indicator of hygienic slaughterhouse practices. Coliforms/Enterobacteriaceae may be useful for trend analysis.

Microorganism	GMP (end of manufacture)	Maximum (end of shelf life)
ACC	$<10^5$	10^7
E. coli	$<10^2$	10^3
Yeasts (sausages & marinated products)	$<10^4$	10^5

Note: for minced meat, mechanically separated meat (MSM) and meat preparations, process hygiene criteria have been set in legislation for *E. coli* and these limits differ from those above – see section 7 for more detail (see also section 6.2).

C: RAW FISH & SHELLFISH

Product examples: Raw fish (whole, fillets, pieces), cold smoked, gravadlax, marinated or pickled (acidified) products, bivalve molluscs e.g. mussels, oysters; crustacea e.g. prawns, scombroid fish e.g. tuna, roll mop herrings.

Storage: Frozen or chilled
Use: To be cooked or ready-to-eat

Pathogens & Toxins

Pathogens may be present; in particular *Vibrio* may occur in warm water fish/shellfish. As filter feeders, bivalve molluscs concentrate contaminants from the water in which they grow, including bacterial pathogens, viruses and algal toxins. Monitoring the incidence of bacterial pathogens and algal toxins may be necessary (see sections 4 and 7 for more detail). Presence of histamine (scombrotoxin) in scombroid fish should also be considered.

Product	Microorganism/Toxin	GMP (end of manufacture)	Maximum (end of shelf life)
To be cooked	Bacterial pathogens	Criteria for absence not generally applicable.	
	Histamine (scombroid fish)	<200mg/kg	400mg/kg
Ready-to-eat	*Salmonella* spp.	ND in 25g	ND in 25g
	V. parahaemolyticus (warm water fish)	ND in 25g	10^2
	L. monocytogenes	ND in 25g	10^2
	Coagulase-positive staphylococci (cold smoked fish)	$<10^2$	10^2
	Histamine (scombroid fish)	<200mg/kg	400mg/kg

Indicators & spoilage microorganisms

ACC is often high in raw products; monitoring may be useful for trend analysis. *E. coli* is commonly used as an indicator of faecal contamination in shellfish.

Product	Microorganism	GMP (end of manufacture)	Maximum (end of shelf life)
To be cooked	*E. coli*	<10	10^2
	Yeasts (pickled/marinated)	$<10^4$	10^5
Ready-to-eat	ACC	$<10^3$	10^4
	E. coli	<10	10^2

D: DOUGHS, PASTA & BATTERS

Product examples: Bread dough, pastry, part-baked products, pasta e.g. spaghetti, tagliatelle, noodles.

Storage: Frozen or chilled
Use: To be cooked

Pathogens

Salmonella spp. and *L. monocytogenes* may be present occasionally in raw products but usually only at low levels and should be killed by cooking. *B. cereus* spores and/or emetic toxin and staphylococcal enterotoxins can survive cooking (>F0 3).

Microorganism	GMP (end of manufacture)	Maximum (end of shelf life)
B. cereus	$<10^2$	10^4
Coagulase positive staphylococci (pasta and batters)	$<10^2$	10^4

Indicators & spoilage microorganisms

ACC and Enterobacteriaceae levels may be very high and variable. *E. coli* may be used to indicate hygiene of production. Lactic acid bacteria may be used to indicate spoilage (except for sourdoughs).

Microorganism	GMP (end of manufacture)	Maximum (end of shelf life)
E. coli	$<10^2$	10^3
Lactic acid bacteria (doughs/pastry)	$<10^5$	10^7

E: FRUIT AND FRUIT JUICES

Product examples: Raw sliced, or chopped fruits and fruit juice, freshly squeezed juices, pasteurised juices and heat processed fruit drinks, fruit cocktail and soft whole fruit.

Storage: Frozen, chilled or ambient
Use: To be cooked or ready-to-eat

Pathogens & Toxins
Pathogens, including parasites and viruses, may be present but are not common. Growth is unlikely in an acidic environment although *Salmonella* spp. and STEC have been found to survive.

Product	Microorganism	GMP (end of manufacture)	Maximum (end of shelf life)
Fruit juice and pre-cut fruit	*Salmonella* spp.	ND in 25g	ND in 25g
	L. monocytogenes	ND in 25g	10^2
	STEC	ND in 25g	ND in 25g

Indicators & spoilage microorganisms
ACC could be high in raw products. Yeast & mould levels are important in relation to spoilage and visual inspection/monitoring of product is particularly relevant. If fruit is to be used for juices the level of yeasts should be low to obtain a realistic shelf-life. *E. coli* may be useful as an indicator of poor agricultural hygiene in harvesting and processing. Growing concern with Alicyclobacillus warrants its assessment in such products.

Product	Microorganism	GMP (end of manufacture)	Maximum (end of shelf life)
Pre-cut fruit (non-heat processed fruit)	Yeast	$<10^3$	10^5
	E. coli	<10	10^2
Fruit juice (unpasteurised)	Yeasts	$<10^3$	10^5
	E. coli	<10	10^2
Fruit juice (pasteurised)	Yeasts	<10	-

F: RAW AND PREPARED VEGETABLES (INCLUDING SALAD VEGETABLES)

Product examples: Raw and prepared vegetables, including salad vegetables and stir-fry, blanched vegetables, frozen vegetables, fresh herbs, bean sprouts, sandwich fillings.

Storage: Frozen, chilled or ambient
Use: To be cooked or ready-to-eat

Pathogens

Pathogens e.g. *Salmonella* spp., *Listeria monocytogenes*, STEC, may be present on raw vegetables, bean sprouts and other sprouted seeds, herbs (fresh cut or growing) and salad vegetables. Washing procedures are unlikely to eliminate them. However, only low incidence and levels may be expected.

Use	Microorganism	GMP (end of manufacture)	Maximum (end of shelf life)
To be washed or cooked	Bacterial pathogens	Criteria for absence not generally applicable however, monitoring incidence may be useful for trend analysis	
Prepared (ready-to-eat)	*Salmonella* spp.	ND in 25g	ND in 25g
	L. monocytogenes	ND in 25g	10^2
	STEC	ND in 25g	ND in 25g

Indicators & spoilage microorganisms

High levels of ACC and Enterobacteriaceae/coliforms are probable and could be derived from the soil or poor handling. *E. coli* levels may be used to monitor the quality of irrigation water and hygiene of handling. If vegetables are to be used in further processing e.g. canning, monitoring for bacterial spore levels may be useful. Visual inspection is particularly important for bacterial rots and mould contamination.

Use	Microorganism	GMP (end of manufacture)	Maximum (end of shelf life)
To be washed or cooked	Criteria not generally applicable		
Prepared (ready-to-eat)	*E. coli*	$<10^2$	10^3

G: PASTEURISED MILK & MILK PRODUCTS

Product examples: Liquid milk, cream, cheese, ice cream, butter, dairy desserts, yoghurts and other fermented products. For dried baby foods see Table O.

Storage: Frozen or chilled
Use: Ready-to-eat

Pathogens

Vegetative pathogens will be absent in properly pasteurised milk and cream; testing for *Salmonella* and *Listeria monocytogenes* is not conducted routinely in these products. Low levels of spores and other thermotolerant microorganisms may occasionally be present.

Microorganism	GMP (end of manufacture)	Maximum (end of shelf life)
Salmonella spp.	ND in 25ml/g	ND in 25ml/g
L. monocytogenes	ND in 25ml/g	10^2
Coagulase-positive staphylococci (Un-ripened Soft Cheeses only)	<10	10^2
B. cereus	$<10^2$	10^3
Staphylococcal enterotoxins (Cheese only)	ND in 25ml/g	-

Indicators & spoilage microorganisms

Criteria applicable depend on the level of heat treatment applied.
Coliforms/Enterobacteriaceae and *E. coli* are most commonly used as indicators of the quality and hygienic processing of dairy products. ACC may be useful for trend analysis but is not applicable to live fermented products.

Product	Microorganism	GMP (end of manufacture)	Maximum (end of shelf life)
Cheese	*E. coli*	$<10^2$	10^2
Liquid milk	Aerobic colony count	$<10^2$	10^4
	Enterobacteriaceae	<1	10

H: NON-PASTEURISED MILK & MILK PRODUCTS

Product examples: Liquid milk, cream, cheese, ice cream, butter, dairy desserts, yoghurts and other fermented dairy products.

Storage: Frozen or chilled
Use: Ready-to-eat

Pathogens
Salmonella spp., *Campylobacter*, STEC, *Listeria monocytogenes* and *Staph. aureus* will occasionally be present in raw milk. Other pathogens and parasites may also occur.

Microorganism	GMP (end of manufacture)	Maximum (end of shelf life)
Salmonella spp.	ND in 25ml/g	ND in 25ml/g
L. monocytogenes	ND in 25ml/g	10^2
Coagulase-positive staphylococci	$<10^2$	10^3
B. cereus	$<10^2$	10^3
STEC	ND in 25ml/g	-
Staphylococcal enterotoxins (Cheese only)	ND in 25g	-

Indicators & spoilage microorganisms

Product	Microorganism	GMP (end of manufacture)	Maximum (end of shelf life)
Soft cheese, Butter	*E. coli*	<10	10^2
Milk	Aerobic colony count	$<10^4$	10^4
	Coliforms	$<10^2$	10^3
Ice cream	Enterobacteriaceae	<10	10^2

I: UHT MILK, CREAM & DAIRY PRODUCTS

Product examples: Long life UHT milks including flavoured products, milkshakes, sterilised milk

Storage: Ambient
Use: Ready-to-eat

Pathogens
Pathogens should be absent.

Microorganism	GMP (end of manufacture)	Maximum (end of shelf life)
Bacterial pathogens	Criteria for absence not generally applicable	

Indicators & spoilage microorganisms
UHT products should be microbiologically stable. Control of the process is essential. Microbiological testing of the finished product is not considered appropriate. Vegetative cells are not expected to be present in these products. Incubation of finished packs followed by examination for signs of microbial growth (pH change, organoleptic defects, blown packs) is most commonly used in industry. Direct inoculation of agar with incubated product or alternative methods such as impedance/conductance and ATP-bioluminescence are also used to detect growth.

Microorganism	GMP (end of manufacture)	Maximum (end of shelf life)
ACC (after incubation of pack)	No evidence of growth	No evidence of growth

J: PART COOKED FOODS

Product examples: Prepared ready meals, pies, and pizzas (meat, fish and or vegetables); coated shellfish, fish and meat products; fresh filled pasta products; egg – liquid.

Storage: Frozen or chilled
Use: To be cooked

Pathogens
Salmonella spp., *Campylobacter* spp., STEC and *L. monocytogenes* could be present (particularly from raw meats, poultry and fish). Spores of *C. perfringens* and *B. cereus* may also be present and monitoring for trend analysis may be useful. Monitoring for *Staph. aureus* should be considered for products receiving significant handling during processing e.g. pizzas and pasta-based products.

Microorganism	GMP (end of manufacture)	Maximum (end of shelf life)
Bacterial pathogens	Criteria for absence not generally applicable Monitoring incidence may be useful for trend analysis.	

Indicators & spoilage microorganisms
Meals of different components, some perhaps partly cooked, cannot easily be regarded as a single sample likely to show a consistent microbial flora. Individual components of meals may merit separate consideration, where *E. coli* and Enterobacteriaceae may be used to indicate hygiene of production.

Microorganism	GMP (end of manufacture)	Maximum (end of shelf life)
Enterobacteriaceae	$<10^2$	10^3
E. coli	<10	10^3

K: CHILLED & FROZEN PROCESSED FOODS

Product examples: Ready-meals; cooked vegetables, pulses, rice, meats, shellfish and fish products; pies, pasties, quiches, flans; sandwiches; sous vide products; fermented, smoked and cured meats; cured and hot smoked fish; sauces; desserts.
Storage: Frozen or chilled

Use: Ready-to-eat or, to be reheated

Pathogens & Toxins
Pathogens may be present in some raw fermented meat products. Spores of psychrotrophic strains of Clostridium botulinum may present a hazard in products given an extended shelf-life at refrigeration temperatures and/or those packed under reduced oxygen conditions and with few, or no, other intrinsic factors present to inhibit bacterial growth. Monitoring the incidence of algal toxins (see sections 4 and 7 for more detail) may be necessary in molluscs and the presence of histamine (scombrotoxin) in scombroid fish should also be considered.

Microorganism	GMP (end of manufacture)	Maximum (end of shelf life)
Salmonella spp.	ND in 25g	ND in 25g
L. monocytogenes (ready-to-eat foods)	ND in 25g	10^2
C. perfringens (cooked meat, vegetables and pulses)	$<10^2$	10^3
B. cereus (rice products, products containing spices and bakery products)	$<10^2$	10^4
Vibrio parahaemolyticus (warm water fish)	ND in 25g	10^2
STEC (ready-to-eat foods)	ND in 25g	ND in 25g
Histamine (scombroid fish)	<200mg/kg	400mg/kg

Indicators & spoilage microorganisms
ACC varies greatly between products depending on processing and stage of shelf-life. Most non-fermented products will meet a limit of 10^5 cfu/g at point of production if hygienically prepared; many will easily meet 10^4 cfu/g. Monitoring of ACC may be useful for trend analysis but is not appropriate for live fermented products. Enterobacteriaceae are a useful monitor of hygiene practice. For some cooked sliced meats, the ACC can reach 10^8 during shelf-life due to the presence and growth of lactic acid bacteria without detrimental effect to the product. Presence of other microorganisms in high numbers ($>10^7$cfu/g) may cause spoilage. Care is therefore required in the use and interpretation of ACC levels.

Microorganism	GMP (end of manufacture)	Maximum (end of shelf life)
ACC (Cooked meats)	$<10^4$	Variable depending on product
Enterobacteriaceae	$<10^2$	10^4

L: BAKERY PRODUCTS

Product examples: Bakery products: bread, crumpets, muffins.

Storage: Ambient

Use: Ready-to-eat or, to be reheated

Pathogens & Toxins

There is a long history of no concerns with respect to proteolytic *C. botulinum* in bakery products although the controlling mechanism is unknown. Long life MAP products should be risk assessed and some level of routine verification testing undertaken at start and end of life to check for no growth for this organism type

Microorganism	GMP (end of manufacture)	Maximum (end of shelf life)
Salmonella spp.	ND in 25g	ND in 25g
B. cereus	$<10^2$	10^4
Coagulase-positive staphylococci	<20	10^4

Indicators & spoilage microorganisms

Taints are often the result of yeast and/or lactic acid bacteria growth. Rope formation (stringy cheese like effect) is caused by Bacillus growth and should be confirmed if found.

Microorganism	GMP (end of manufacture)	Maximum (end of shelf life)
Enterobacteriaceae	<10	-
Yeasts & moulds	Acceptable visually and organoleptically	-

M: DRIED FOODS, TO BE COOKED

Product examples: Rice, pulses, cereals, flour, dried vegetables, dried fruit, dried coconut, pasta/noodles, dry mixes, dried herbs and spices.

Storage: Ambient
Use: To be cooked

Pathogens & Toxins

Pathogens may be present; monitoring may be useful for trend analysis. Spores may survive cooking. Heat stable staphylococcal enterotoxins and Bacillus spp. may not be inactivated by cooking. The presence of mycotoxins should be considered.

Microorganism	GMP (end of manufacture)	Maximum (end of shelf life)
Staph. aureus (pasta) (pre- and post-heating stage)	$<10^2$	10^4
B. cereus	$<10^2$	10^4
C. perfringens	$<10^2$	10^3

Indicators & spoilage microorganisms

ACC and levels of Enterobacteriaceae and spore-forming bacteria will vary according to commodity; monitoring may be of value for trend analysis. Bacterial spore levels should be low for some uses e.g. canning.

Microorganism	GMP (end of manufacture)	Maximum (end of shelf life)
E. coli	$<10^2$	10^3
Yeast (fruit)	$<10^3$	10^5
Moulds	$<10^3$	10^5

N: DRIED FOODS, READY TO EAT

Product examples: Processed nuts, fruits, coconut, cocoa powder, breakfast cereals, crisps, snacks, confectionery, chocolate, filled or topped biscuits and cakes; decontaminated herbs, spices and flavourings.

Storage: Ambient
Use: Ready-to-eat

Pathogens & Toxins
Pathogens may be present; bacterial spore-formers are common in herbs and spices. Aflatoxin testing may be relevant for nut and nut-containing products.

Microorganism	GMP (end of manufacture)	Maximum (end of shelf life)
Salmonella spp.	ND in 25g	ND in 25g
Coagulase-positive staphylococci	<20	10^3
B. cereus (herbs and spices)	$<10^2$	10^3
C. perfringens (herbs and spices)	$<10^2$	10^3

Indicators & spoilage microorganisms
ACC and levels of Enterobacteriaceae may be as high as 10^7cfu/g in herbs and spices. Monitoring may be useful for trend analysis of individual products.

Microorganism	GMP (end of manufacture)	Maximum (end of shelf life)
E. coli (raw dried foods, ready-to-eat)	<10	10^2
Enterobacteriaceae (processed)	<10	10^2
Yeasts (fruits)	$<10^3$	10^5
Moulds (raw)	$<10^3$	10^4

O: DRIED HEAT PROCESSED FOODS, READY-TO-EAT AFTER REHYDRATION

Product examples: Soup mixes, gravy mixes, sauce mixes, dessert mixes, egg – dried, milk powder (not including baby foods or infant formulae; see Section 7), pot snacks, breakfast cereals.

Storage: Ambient
Use: Ready-to-eat after rehydration

Pathogens
These products may contain bacterial spore-formers. If reconstituted product is held at warm temperatures (30-50°C) then pathogens such as *C. perfringens* may grow to levels causing illness. This must be considered when determining appropriate criteria as more stringent limits than those indicated below may be applicable.

Microorganism	GMP (end of manufacture)	Maximum (end of shelf life)
Salmonella spp.	ND in 25g	ND in 25g
S. aureus	<20	10^3
B. cereus	$<10^2$	10^3
C. perfringens	$<10^2$	10^3

Indicators & spoilage microorganisms
For some foods, monitoring the ACC may be useful for trend analysis but it may be high in products containing herbs and spices.

Microorganism	GMP (end of manufacture)	Maximum (end of shelf life)
Enterobacteriaceae	<10	10^2
E. coli	<10	10^2

P: DRIED BABY FOODS

Product examples:	Dry mixes, milk powder.
Storage:	Ambient
Use:	Ready-to-eat after rehydration

Pathogens

Because babies are more susceptible to food poisoning than adults, these products are normally made to very stringent criteria. Low microbial limits are applied together with an increased level of sampling. If reconstituted product is held at warm temperatures (30-50°C) then pathogens may grow to levels causing illness. This must be considered when determining appropriate criteria as even more stringent limits than those indicated below may be applicable (see also Section 7).

Microorganism	GMP (end of manufacture)	Maximum (end of shelf life)
Salmonella spp.	ND in 7500g (usually individual 25g samples examined)	ND in 750g
L. monocytogenes	ND in 25g	ND in 25g
C. perfringens	<10	10^2
B. cereus	<10	10^2
Coagulase-positive staphylococci	10^2	10^2
Cronobacter spp.	ND in 300g	ND in 300g

Indicators & spoilage microorganisms

ACC and Enterobacteriaceae may be useful indicators of process hygiene. Stricter legislative criteria have been set for infant formulae – see section 7.

Microorganism	GMP (end of manufacture)	Maximum (end of shelf life)
ACC	$<10^3$	10^4
Enterobacteriaceae	ND in 100g	ND in 100g
E. coli	ND in 1g	10

Q: CANNED, POUCHED OR BOTTLED FOODS (>$F_0$3 PROCESS)

Product examples: Canned meats, fish, vegetables and soups, ready-meals in pouches, baby foods in jars.

Storage: Ambient
Use: Any

Pathogens & Toxins
Bacterial spores may survive the heat process. Control of processing is paramount including can seam integrity, post-heat treatment cooling and container handling hygiene. Microbiological monitoring of processed product is not recommended. Histamine (scombrotoxin) testing may be relevant for scombroid fish products.

Microorganism	GMP (end of manufacture)	Maximum (end of shelf life)
Bacterial pathogens	Criteria for absence not generally applicable	
Histamine (scombroid fish)	<200mg/kg	400mg/kg

Indicators & spoilage microorganisms
Bacterial spores may survive processing. Control of the process is essential. Microbiological testing of the finished product is not considered appropriate. Incubation of finished packs followed by examination for signs of microbial growth or spoilage (pH change, organoleptic defects, blown packs) is most commonly used in industry.

Microorganism	GMP (end of manufacture)	Maximum (end of shelf life)
All	No evidence of growth (after incubation of pack)	

R: PRESERVED FOODS - HEAT TREATED (<$F_0$3 PROCESS), INTERMEDIATE MOISTURE FOOD (IMF) OR LOW pH

Product examples: Salami, dry cured hams, bresaola, pastrami, vegetables and herbs in oil, canned or bottled fruits, jams, relishes, pickles, sauces.

Storage: Ambient
Use: Ready-to-eat

Pathogens & Toxins

Bacterial spores may be present from raw materials e.g. vegetables and herbs. Control of product physico-chemical characteristics e.g. pH, water activity during the process and maintenance of these conditions throughout product shelf-life are paramount for ensuring safety. Histamine (scombrotoxin) testing may be relevant for scombroid fish products.

Microorganism	GMP (end of manufacture)	Maximum (end of shelf life)
Salmonella spp. (not heat treated e.g. herbs in oil)	ND in 25g	ND in 25g
L. monocytogenes	ND in 25g	10^2
C. perfringens	$<10^2$	10^3
Coagulase-positive staphylococci (meat products)	<20	10^3
STEC (fermented and dried meats)	ND in 25g	ND in 25g
Histamine (scombroid fish)	<200mg/kg	400mg/kg

Indicators & spoilage microorganisms

ACC and Enterobacteriaceae may be useful for monitoring process hygiene and spoilage potential. ACC is not applicable to live, fermented products. Yeasts and moulds (see Appendix I - 'Yeasts and Moulds') may also be significant in the spoilage of some products.

Microorganism	GMP (end of manufacture)	Maximum (end of shelf life)
Enterobacteriaceae (not heat processed)	$<10^2$	10^4
Enterobacteriaceae (heat processed)	<10	10^2
E. coli (not heat processed)	<10	10^2
E. coli (heat processed)	<10	10
Yeasts	$<10^3$	10^5
Moulds (not mould ripened)	$<10^2$	10^4

S: NON-DAIRY FATS, OILS & SPREADS

Product examples: Cooking oils, fat spreads, margarines, lard, virgin olive oil, nut oils, nut butters, cocoa butter.

Storage: Ambient or chilled
Use: Any

Pathogens & Toxins

Pathogens should be absent in hot refined oils and 100% fat products packed under good hygienic conditions and monitoring is not generally required. Control of product physico-chemical properties (emulsion characteristics, salt, pH) and hygienic processing conditions is important for ensuring the safety of fat spreads and margarines. Monitoring may be required for some cold pressed products. Mould contaminating nuts may produce mycotoxins which can carry through to the finished product. Raw material nuts should be routinely screened for mycotoxin contamination.

Microorganism	GMP (end of manufacture)	Maximum (end of shelf life)
Bacterial pathogens (heat treated)	Criteria for absence not generally applicable	
Salmonella spp. (not heat treated)	ND in 25g	ND in 25g
L. monocytogenes (chilled)	ND in 25g	10^2

Indicators & spoilage microorganisms

ACC and Enterobacteriaceae are normally low and may be useful for trend analysis for monitoring process hygiene. Lipolytic microorganisms may cause rancidity in fat spreads and testing for these microorganisms may be appropriate. Control of moisture during processing, packing and shelf-life is important for solid fat products as the presence of moisture can allow spoilage by surface mould growth.

Microorganism	GMP (end of manufacture)	Maximum (end of shelf life)
Enterobacteriaceae (fat spreads)	<10	10^3

T: SOFT DRINKS AND ALCOHOLIC BEVERAGES

Product examples: Fruit drinks (heat processed and preserved), beers, wines, colas, lemonade.

Storage: Ambient
Use: Ready-to-drink

Pathogens
Pathogens are unlikely to be present but could survive and grow in low or no alcohol products.

Microorganism	GMP (end of manufacture)	Maximum (end of shelf life)
Bacterial pathogens	Criteria for absence not generally applicable	

Indicators & spoilage microorganisms
Raw material quality, processing hygiene and control of the product formulation are important in spoilage prevention for these products. Yeasts, moulds and lactic acid bacteria are the main causes of spoilage. Incubation of finished packs followed by visual examination for signs of microbial spoilage e.g. haze, sediment or gas is usually sufficient.

Microorganism	GMP (end of manufacture)	Maximum (end of shelf life)
Yeasts (filtered or heat processed)	ND in final pack	Evidence of growth
Lactic acid bacteria (filtered or heat processed)	ND in final pack	Evidence of growth
Yeasts (not filtered or heat processed)	<10	Evidence of growth
Lactic acid bacteria (not filtered or heat processed)	<10	Evidence of growth

OTHER STATUTORY CRITERIA

Most microbiological criteria set in EU law are contained in Commission Regulation (EC) No. 2073/2005 on microbiological criteria for foodstuffs (as amended). The regulation ensures that criteria are consistent, based on up-to-date scientific risk assessment and relevant to consumer health protection. It also introduces criteria for certain food groups and microorganisms not previously included in Community legislation (for example criteria for pre-cut fruit and vegetables, powdered infant formulae, *Salmonella* on meat carcasses and *L. monocytogenes* in ready-to-eat food products). The Regulation is open to review and criteria may be amended, as microbiology and food safety issues develop over time, for example a new criteria was introduced for STEC in sprouted seeds following the outbreak of *E. coli* O104:H4 in fenugreek seeds.

Regulation (EC) 2073/2005 establishes two types of criteria and requires that food business operators take corrective action when these criteria are not met [Art. 3(1)]. These two types are:

> **Food Safety Criteria (FSC)** which should be used to assess the safety of a product or batch of foodstuffs by monitoring pathogens; and
>
> **Process Hygiene Criteria (PHC)** which should be used to help show the production processes are operating properly by monitoring indicator organisms (and some pathogenic organisms such as *Salmonella* on meat carcases).

FSC apply throughout the product shelf-life and if tests indicate that the criteria are not met, the food business operator will not be able to place the food on the market or will need to remove it from the market. In some cases, a product recall may be required. The food safety management procedures should also be reviewed to ensure the products comply in the future.

PHC apply throughout processing, manufacturing and handling up until food is placed on the market. If a process criterion is exceeded this should lead to a review of current procedures to improve production hygiene. It may also prompt further testing e.g. on detection of coagulase-positive staphylococci in certain dairy products, toxin testing must be undertaken.

The Regulation applies to all foods that are placed on the market in the EU, whether they are produced in an EU Member State or a third country. Businesses exporting to non-EU countries will also need to comply with relevant national microbiological criteria (not listed here).

Businesses must also take account of microbiological hazards other than those specified in the Regulation where necessary to fulfil their general obligation not to place unsafe food on the market, as required under Commission Regulation (EC) 178/2002 which establishes the general principles and requirements of food and feed law. Food business operators are required to inform their Local Authority of any unacceptable test results. In addition, to help monitor and collect data on the occurrence of zoonoses in food, as required by Directive 2003/99/EC, businesses should retain relevant isolates and send them to the appropriate Reference Laboratory for characterisation and typing. Directive 2003/99/EC lists the relevant zoonotic agents of concern.

Whilst the Commission Regulation focuses mostly on raw and finished products, it also specifies certain requirements for sampling processing areas and food production equipment. For example, it requires that ISO standard 18593 is used as a reference method. It also requires that food business operators manufacturing RTE foods that may pose a *L. monocytogenes* risk shall sample processing areas and equipment for *L. monocytogenes* as part of their sampling scheme and that operators manufacturing dried infant formulae or dried foods for special medical purposes intended for infants <6 months of age, which pose a *Cronobacter* spp. risk shall monitor processing areas and equipment for Enterobacteriaceae.

Considering raw and finished goods, the following tables provide an overview of information on microbiological criteria and non-microbiological hygiene criteria in the current EU and UK legislation.

To ensure the correct context for application, the criteria should be read in conjunction with the appropriate latest amended versions of legislation.

Definitions:

m = threshold below which all results are considered satisfactory

M = acceptability threshold, the result is considered unsatisfactory if one or more units give values ≥ M

n = number of units making up the sample

c = number of units in the sample giving values between m and M; the sample being considered acceptable if the values of the other sample units are ≤ m

6.5. LEGISLATION: COMMISSION REGULATION (EC) NO. 2073/2005 ON MICROBIOLOGICAL CRITERIA FOR FOODSTUFFS AS AMENDED

FOOD CATEGORY	MICROORGANISM	LEVEL (per g or ml, unless	STATUS/ACTION
Ready-to-eat foods intended for infants and ready-to-eat foods for special medical purposes	*L. monocytogenes* (FSC)	M = Absent in 25g , n = 10, c = 0	Withdraw from the market and, if necessary, recall in accordance with requirements of Article 19 178/2002. Additionally, other corrective actions defined in operators' HACCP-based procedures must be carried out, including modifications to the HACCP-based procedures to prevent recurrence of the hazard.
Ready-to-eat foods able to support the growth of *L. monocytogenes*, other than those intended for infants and for special medical purposes	*L. monocytogenes* (FSC)	M = 10^2 n = 5, c = 0 Or M = Absent in 25g, n = 5, c = 0	Withdraw from the market and, if necessary, recall in accordance with requirements of Article 19 178/2002. Additionally, other corrective actions defined in operators' HACCP-based procedures must be carried out, including modifications to the HACCP-based procedures to prevent recurrence of the hazard.
Ready-to-eat foods unable to support the growth of *L. monocytogenes*, other than those intended for infants and for special medical purposes	*L. monocytogenes* (FSC)	M = 100, n = 5, c = 0	Withdraw from the market and, if necessary, recall in accordance with requirements of Article 19 178/2002. Additionally, other corrective actions defined in operators' HACCP-based procedures must be carried out, including modifications to the HACCP-based procedures to prevent recurrence of the hazard.

FOOD CATEGORY	MICROORGANISM	LEVEL (per g or ml, unless otherwise specified)	STATUS/ACTION
Dried infant formulae and dried dietary foods for special medical purposes intended for infants below six months of age	*Salmonella* (FSC)	M = Absent in 25g, n = 30, c = 0	Withdraw from market and, if necessary, recall in accordance with requirements of Article 19 178/2002. Additionally, other corrective actions defined in operators' HACCP-based procedures must be carried out, including modifications to HACCP-based procedures to prevent hazard recurrence
Dried infant formulae and dried dietary foods for special medical purposes intended for infants below six months of age	*Cronobacter* spp. (FSC)	M = Absent in 10g, n = 30, c = 0	Withdraw from market and, if necessary, recall in accordance with requirements of Article 19 178/2002. Additionally, other corrective actions defined in operators' HACCP-based procedures must be carried out, including modifications to HACCP-based procedures to prevent hazard recurrence
Dried follow-on formulae	*Salmonella* (FSC)	M = Absent in 25g, n = 30, c = 0	Withdraw from market and, if necessary, recall in accordance with requirements of Article 19 178/2002. Additionally, other corrective actions defined in operators' HACCP-based procedures must be carried out, including modifications to HACCP-based procedures to prevent hazard recurrence

FOOD CATEGORY	MICROORGANISM	LEVEL (per g or ml, unless otherwise specified)	STATUS/ACTION
Dried infant formulae and dried dietary foods for special medical purposes intended for infants below six months of age	Enterobacteriaceae (PHC)	M = Absent in 10g, n = 10, c = 0	Improvements in production hygiene to minimise contamination. Note: Parallel testing for Enterobacteriaceae and *Cronobacter* spp. (formerly *E. sakazakii*) shall be conducted unless a correlation between these microorganisms has been established on an individual plant level. If Enterobacteriaceae are detected in any of the product samples tested in such a plant, the batch must be tested for *E. sakazakii*.
Dried infant formulae and dried dietary foods for special medical purposes intended for infants below six months of age	Presumptive *Bacillus cereus* (PHC)	m = 50, M = 500, n = 5, c = 1	Improvements in production hygiene, preventing recontamination & selection of raw material.
Dried follow-on formulae	Enterobacteriaceae (PHC)	M = Absent in 10g, n = 5, c = 0	Improvements in production hygiene to minimise contamination.

FOOD CATEGORY	MICROORGANISM	LEVEL (per g or ml, unless otherwise specified)	STATUS/ACTION
Minced meat and meat preparations intended to be eaten raw	*Salmonella* (FSC)	M = Absent in 25g, n = 5, c = 0	Withdraw from the market and, if necessary, recall in accordance with requirements of Article 19 178/2002. Additionally, other corrective actions defined in operators' HACCP-based procedures must be carried out, including modifications to the HACCP-based procedures to prevent recurrence of the hazard.
Minced meat and meat preparations made from poultry meat intended to be eaten cooked	*Salmonella* (FSC)	M = Absent in 25g, n = 5, c = 0	Withdraw from the market and, if necessary, recall in accordance with requirements of Article 19 178/2002. Additionally, other corrective actions defined in operators' HACCP-based procedures must be carried out, including modifications to the HACCP-based procedures to prevent recurrence of the hazard.
Minced meat and meat preparations made from other species than poultry intended to be eaten cooked	*Salmonella* (FSC)	M = Absent in 10g, n = 5, c = 0 (For intra-community trade) Or M = Absent in 10g, n = 5, c = 1	Withdraw from the market and, if necessary, recall in accordance with requirements of Article 19 178/2002. Additionally, other corrective actions defined in operators' HACCP-based procedures must be carried out, including modifications to the HACCP-based procedures to prevent recurrence of the hazard.

FOOD CATEGORY	MICROORGANISM	LEVEL (per g or ml, unless otherwise specified)	STATUS/ACTION
Mechanically separated meat (MSM)	*Salmonella* (FSC)	M = Absent in 10g, n = 5, c = 0	Withdraw from the market and, if necessary, recall in accordance with requirements of Article 19 178/2002. Additionally, other corrective actions defined in operators' HACCP-based procedures must be carried out, including modifications to the HACCP-based procedures to prevent recurrence of the hazard.
Meat products intended to be eaten raw (excluding products where the manufacturing process or the composition of the product will eliminate the *Salmonella* risk)	*Salmonella* (FSC)	M = Absent in 25g, n = 5, c = 0	Withdraw from the market and, if necessary, recall in accordance with requirements of Article 19 178/2002. Additionally, other corrective actions defined in operators' HACCP-based procedures must be carried out, including modifications to the HACCP-based procedures to prevent recurrence of the hazard.
Meat products made from poultry meat intended to be eaten cooked	*Salmonella* (FSC)	M = Absent in 25g, n = 5, c = 0	Withdraw from the market and, if necessary, recall in accordance with requirements of Article 19 178/2002. Additionally, other corrective actions defined in operators' HACCP-based procedures must be carried out, including modifications to the HACCP-based procedures to prevent recurrence of the hazard.

FOOD CATEGORY	MICROORGANISM	LEVEL (per g or ml, unless otherwise specified)	STATUS/ACTION
Gelatine and collagen	*Salmonella* (FSC)	M = Absent in 25g, n = 5, c = 0	Withdraw from the market and, if necessary, recall in accordance with requirements of Article 19 178/2002. Additionally, other corrective actions defined in operators' HACCP-based procedures must be carried out, including modifications to the HACCP-based procedures to prevent recurrence of the hazard.
Minced meat	Aerobic colony count (PHC)	$m = 5x10^5$, $M = 5x10^6$, n = 5, c = 2	Improvements in production hygiene and improvements in selection and/or origin of raw materials
Minced meat	*E. coli* (PHC)	m = 50, M = 500, n = 5, c = 2	Improvements in production hygiene and improvements in selection and/or origin of raw materials
Mechanically separated meat (MSM)	Aerobic colony count (PHC)	$m = 5x10^5$, $M = 5x10^6$, n = 5, c = 2	Improvements in production hygiene and improvements in selection and/or origin of raw materials

FOOD CATEGORY	MICROORGANISM	LEVEL (per g or ml, unless otherwise specified)	STATUS/ACTION
Mechanically separated meat (MSM)	*E. coli* (PHC)	m = 50, M = $5x10^2$, n = 5, c = 2	Improvements in production hygiene and improvements in selection and/or origin of raw materials
Meat preparations	*E. coli* (PHC)	m = $5x10^2$, M = $5x10^3$, n = 5, c = 2	Improvements in production hygiene and improvements in selection and/or origin of raw materials
Carcasses of cattle, sheep, goats and horses	Aerobic colony count (PHC)	m = 3.5 log cfu/cm2 daily mean log, M = 5.0 log cfu/cm2 daily mean log	Improvements in slaughter hygiene and review of process controls
Carcasses of cattle, sheep, goats and horses	Enterobacteriaceae (PHC)	m = 1.5 log cfu/cm2 daily mean log, M = 2.5 log cfu/cm2 daily mean log	Improvements in slaughter hygiene and review of process controls
Carcasses of cattle, sheep, goats and horses	*Salmonella* (PHC)	M = Absent in the area tested per carcass, n = 50, c = 2	Improvements in slaughter hygiene and review of process controls

FOOD CATEGORY	MICROORGANISM	LEVEL (per g or ml, unless otherwise specified)	STATUS/ACTION
Carcasses of pigs	Aerobic colony count (PHC)	m = 4.0 log cfu/cm2 daily mean log, M =5.0 log cfu/cm2 daily mean log	Improvements in slaughter hygiene and review of process controls
Carcasses of pigs	Enterobacteriaceae (PHC)	m = 2.0 log cfu/cm2 daily mean log, M = 3.0 log cfu/cm2 daily mean log	Improvements in slaughter hygiene and review of process controls
Carcasses of pigs	*Salmonella* (PHC)	M = Absent in the area tested per carcass, n = 50, c = 3	Improvements in slaughter hygiene and review of process controls

FOOD CATEGORY	MICROORGANISM	LEVEL (per g or ml, unless otherwise specified)	STATUS/ACTION
Fresh poultry meat	*Salmonella* Typhimurium *Salmonella* Enteritidis (FSC)	Absent in 25g, n = 5, c = 0	Withdraw from the market and, if necessary, recall in accordance with requirements of Article 19 178/2002. Additionally, other corrective actions defined in operators' HACCP-based procedures must be carried out, including modifications to the HACCP-based procedures to prevent recurrence of the hazard
Poultry carcasses of broilers and turkeys	*Salmonella* (PHC)	Absence in 25g of a pooled sample of neck skin, n = 50, c = 7	Improvements in slaughter hygiene and review of process controls
Poultry carcasses of broilers	*Campylobacter* (PHC)	M = 10^3 cfu/g, n =50, c = 20 From 1.1.2020 c=15; From 1.1.2025 c=10	Improvements in slaughter hygiene and review of process controls

FOOD CATEGORY	MICROORGANISM	LEVEL (per g or ml, unless otherwise specified)	STATUS/ACTION
Cooked crustaceans and molluscan shellfish	*Salmonella* (FSC)	M = Absent in 25g, n = 5, c = 0	Withdraw from the market and, if necessary, recall in accordance with requirements of Article 19 178/2002. Additionally, other corrective actions defined in operators' HACCP-based procedures must be carried out, including modifications to HACCP-based procedures to prevent recurrence of the hazard.
Live bivalve molluscs and live echinoderms, tunicates and gastropods	*Salmonella* (FSC)	M = Absent in 25g, n = 5, c = 0	Withdraw from the market and, if necessary, recall in accordance with requirements of Article 19 178/2002. Additionally, other corrective actions defined in operators' HACCP-based procedures must be carried out, including modifications to HACCP-based procedures to prevent recurrence of the hazard.
Live bivalve molluscs and live echinoderms, tunicates and gastropods	*E. coli* (FSC)	230 MPN/100g of flesh and intra-valvular liquid, M = 700 MPN/100g, n=5, c=1	Withdraw from the market and, if necessary, recall in accordance with requirements of Article 19 178/2002. Additionally, other corrective actions defined in operators' HACCP-based procedures must be carried out, including modifications to HACCP-based procedures to prevent recurrence of the hazard.

FOOD CATEGORY	MICROORGANISM	LEVEL (per g or ml, unless otherwise specified)	STATUS/ACTION
Fishery products from fish species associated with a high amount of histidine	Histamine (FSC)	m = 100 mg/kg M = 200 mg/kg n = 9, c = 2	Withdraw from the market and, if necessary, recall in accordance with requirements of Article 19 178/2002. Additionally, other corrective actions defined in operators' HACCP-based procedures must be carried out, including modifications to the HACCP-based procedures to prevent recurrence of the hazard.
Fishery products which have undergone enzyme maturation treatment in brine, manufactured from fish species associated with a high amount of histidine	Histamine (FSC)	m = 200 mg/kg M = 400 mg/kg n = 9, c = 2	Withdraw from the market and, if necessary, recall in accordance with requirements of Article 19 178/2002. Additionally, other corrective actions defined in operators' HACCP-based procedures must be carried out, including modifications to the HACCP-based procedures to prevent recurrence of the hazard.
Fish sauce produced by fermentation of fishery products	Histamine (FSC)	m = 400 mg/kg M = 400 mg/kg n = 1, c = 0	Withdraw from the market and, if necessary, recall in accordance with requirements of Article 19 178/2002. Additionally, other corrective actions defined in operators' HACCP-based procedures must be carried out, including modifications to the HACCP-based procedures to prevent recurrence of the hazard.

FOOD CATEGORY	**MICROORGANISM**	**LEVEL (per g or ml, unless otherwise specified)**	**STATUS/ACTION**
Shelled and shucked products of cooked crustaceans and molluscan shellfish	*E. coli* (PHC)	m = 1, M = 10, n = 5, c = 2	Improvements in slaughter hygiene and review of process controls
Shelled and shucked products of cooked crustaceans and molluscan shellfish	Coagulase-positive staphylococci (PHC)	m = 10^2, M = 10^3, n = 5, c = 2	Improvements in slaughter hygiene and review of process controls

FOOD CATEGORY	MICROORGANISM	LEVEL (per g or ml, unless otherwise specified)	STATUS/ACTION
Milk powder and whey powder	*Salmonella* (FSC)	M = Absent in 25g, n = 5, c = 0	Withdraw from the market and, if necessary, recall in accordance with requirements of Article 19 178/2002. Additionally, other corrective actions defined in operators' HACCP-based procedures must be carried out, including modifications to the HACCP-based procedures to prevent recurrence of the hazard.
Cheeses, milk powder and whey powder (as referred to in the coagulase-positive staphylococci criteria in Chapter 2.2 of Annex 1 2073/2005)	Staphylococcal enterotoxins (FSC)	M = Not detected in 25g, n = 5, c = 0	Withdraw from the market and, if necessary, recall in accordance with requirements of Article 19 178/2002. Additionally, other corrective actions defined in operators' HACCP-based procedures must be carried out, including modifications to the HACCP-based procedures to prevent recurrence of the hazard.
Cheeses, butter and cream made from raw milk or milk that has undergone a lower heat treatment than pasteurisation	*Salmonella* (FSC)	M = Absent in 25g, n = 5, c = 0	Withdraw from the market and, if necessary, recall in accordance with requirements of Article 19 178/2002. Additionally, other corrective actions defined in operators' HACCP-based procedures must be carried out, including modifications to the HACCP-based procedures to prevent recurrence of the hazard.

FOOD CATEGORY	MICROORGANISM	LEVEL (per g or ml, unless otherwise specified)	STATUS/ACTION
Ice cream (containing milk ingredients), excluding products where the manufacturing process or the composition of the product will eliminate the *Salmonella* risk	*Salmonella* (FSC)	M = Absent in 25g, n = 5, c = 0	Withdraw from market and, if necessary, recall in accordance with requirements of Article 19 178/2002. Additionally, other corrective actions defined in operators' HACCP-based procedures must be carried out, including modifications to HACCP-based procedures to prevent hazard recurrence
Milk powder and whey powder	Coagulase-positive staphylococci (PHC)	m = 10, M = 10^2, n = 5, c = 2	Improvements in production hygiene. If values > 100 000 are detected, the batch has to be tested for staphylococcal enterotoxins.
Pasteurised milk and other pasteurised liquid dairy products	Enterobacteriaceae (PHC)	m = <1, M = 5, n = 5, c = 0	Check on the efficiency of heat- treatment and prevention of recontamination.
Milk powder and whey powder	Enterobacteriaceae (PHC)	M = 10, n = 5, c = 0	Check on the efficiency of heat- treatment and prevention of recontamination.

FOOD CATEGORY	MICROORGANISM	LEVEL (per g or ml, unless otherwise specified)	STATUS/ACTION
Cheeses made from milk or whey that has undergone heat treatment	*E. coli* (PHC)	$m = 10^2$, $M = 10^3$, $n = 5$, $c = 2$	Improvements in production hygiene and selection of raw materials.
Cheeses made from raw milk	Coagulase-positive staphylococci (PHC)	$m = 10^4$, $M = 10^5$, $n = 5$, $c = 2$	Improvements in production hygiene and selection of raw materials. If values $>1x10^5$ are detected, cheese batch has to be tested for staph. enterotoxins.
Cheeses made from milk that has undergone lower heat treatment than pasteurisation and ripened cheeses made from milk/whey that has undergone pasteurisation or stronger heat	Coagulase-positive staphylococci (PHC)	$m = 10^2$, $M = 10^3$, $n = 5$, $c = 2$	Improvements in production hygiene and selection of raw materials. If values $>1x10^5$ are detected, cheese batch has to be tested for staph. enterotoxins.
Un-ripened soft cheeses (fresh cheeses) made from milk or whey that has undergone pasteurisation or a stronger heat treatment	Coagulase-positive staphylococci (PHC)	$m = 10$, $M = 10^2$, $n = 5$, $c = 2$	Improvements in production hygiene and selection of raw materials. If values $>1x10^5$ are detected, cheese batch has to be tested for staph. enterotoxins.

FOOD CATEGORY	MICROORGANISM	LEVEL (per g or ml, unless otherwise specified)	STATUS/ACTION
Butter and cream made from raw milk or milk that has undergone a lower heat treatment than pasteurisation	*E. coli* (PHC)	$m = 10$, $M = 10^2$, $n = 5$, $c = 2$	Improvements in production hygiene and selection of raw materials
Ice cream (containing milk ingredients) and frozen dairy desserts	Enterobacteriaceae (PHC)	$m = 10$, $M = 10^2$, $n = 5$, $c = 2$	Improvements in production hygiene.
Egg products, excluding products where the manufacturing process or the composition of the product will eliminate the *Salmonella* risk	*Salmonella* (FSC)	M = Absent in 25g, $n = 5$, $c = 0$	Withdraw from the market and, if necessary, recall in accordance with requirements of Article 19 178/2002. Additionally, other corrective actions defined in operators' HACCP-based procedures must be carried out, including modifications to the HACCP-based procedures to prevent recurence of the hazard.
Ready-to-eat foods containing raw egg, excluding products where the manufacturing process or the composition of the product will eliminate the *Salmonella* risk	*Salmonella* (FSC)	M = Absent in 25g, $n = 5$, $c = 0$	Withdraw from the market and, if necessary, recall in accordance with requirements of Article 19 178/2002. Additionally, other corrective actions defined in operators' HACCP-based procedures must be carried out, including modifications to the HACCP-based procedures to prevent recurence of the hazard.

FOOD CATEGORY	MICROORGANISM	LEVEL (per g or ml, unless otherwise specified)	STATUS/ACTION
Egg products	Enterobacteriaceae (PHC)	m = 10, M = 100, n = 5, c = 2	Checks on the efficiency of the heat treatment and prevention of recontamination.
Sprouts, excluding those that have received a treatment effective to eliminate *Salmonella* and STEC	STEC 0157, 026, 0111, 0103, 0145 and 0104:H4 (FSC)	Absent in 25g, n = 5, c = 0	Withdraw from the market and, if necessary, recall in accordance with requirements of Article 19 178/2002. Additionally, other corrective actions defined in operators' HACCP-based procedures must be carried out, including modifications to the HACCP-based procedures to prevent recurrence of the hazard.
Sprouted seeds (ready-to-eat)	*Salmonella* (FSC)	M = Absent in 25g, n = 5, c = 0	Withdraw from the market and, if necessary, recall in accordance with requirements of Article 19 178/2002. Additionally, other corrective actions defined in operators' HACCP-based procedures must be carried out, including modifications to the HACCP-based procedures to prevent recurrence of the hazard.

FOOD CATEGORY	MICROORGANISM	LEVEL (per g or ml, unless otherwise specified)	STATUS/ACTION
Pre-cut fruit and vegetables (ready-to-eat)	*Salmonella* (FSC)	M = Absent in 25g, n = 5, c = 0	Withdraw from the market and, if necessary, recall in accordance with requirements of Article 19 178/2002. Additionally, other corrective actions defined in operators' HACCP-based procedures must be carried out, including modifications to the HACCP-based procedures to prevent recurrence of the hazard.
Pre-cut fruit and vegetables (ready-to-eat)	*E. coli* (PHC)	m = 100, M = 1000, n = 5, c = 2	Improvements in production hygiene and selection of raw materials
Unpasteurised fruit and vegetable juices (ready-to-eat), unless subjected to a bactericidal process with a similar effect to pasteurisation on *E. coli* and *Salmonella*	*Salmonella* (FSC)	M = Absent in 25g, n = 5, c = 0	Withdraw from the market and, if necessary, recall in accordance with requirements of Article 19 178/2002. Additionally, other corrective actions defined in operators' HACCP-based procedures must be carried out, including modifications to the HACCP-based procedures to prevent recurrence of the hazard.
Unpasteurised fruit and vegetable juices (ready-to-eat), unless subjected to a bactericidal process with a similar effect to pasteurisation on *E. coli* and *Salmonella*	*E. coli* (PHC)	m = 10^2, M = 10^3, n = 5, c = 2	Improvements in production hygiene and selection of raw materials

6.6. LEGISLATION: REGULATION (EC) NO. 853/2004 OF THE EUROPEAN PARLIAMENT AND OF THE COUNCIL LAYING DOWN SPECIFIC HYGIENE RULES FOR FOOD OF ANIMAL ORIGIN AS AMENDED

FOOD CATEGORY	MICROORGANISM	LEVEL (per g or ml, unless otherwise specified)	STATUS/ACTION
Live bivalve molluscs	Paralytic shellfish poison (PSP)	≤ 800 μ g/kg	Withdraw from market
Live bivalve molluscs	Amnesic shellfish poisoning (ASP)	≤ 20 mg domoic acid/kg	Withdraw from market
Live bivalve molluscs	Okadaic acid, dinophysistoxins and pectenotoxins together	≤ 160 μ g okadaic acid equivalents/kg	Withdraw from market
Live bivalve molluscs	Yessotoxins	≤ 3.75 mg yessotoxin equivalent/kg	Withdraw from market
Live bivalve molluscs	Azaspiracids	≤ 160 μ g azaspiracid equivalents/kg	Withdraw from market

FOOD CATEGORY	MICROORGANISM	LEVEL (per g or ml, unless otherwise specified)	STATUS/ACTION
Raw cows' milk at milk production holdings	Plate count (30° C)	$M = \leq 10^5$ (rolling geometric average over a 2-month period, with at least 2 samples per month)	If standards exceeded, inform the competent authority and take steps to correct the situation.
Raw cows' milk at milk production holdings	Somatic cell count (per ml)	$M = \leq 4x10^5$ (rolling geometric average over a 3-month period, with at least 1 sample per month), unless the competent authority specifies another methodology to take account of seasonal variations in production levels	If standards exceeded, inform the competent authority and take steps to correct the situation.
Raw milk from species other than cows at milk production holdings	Plate count (30° C)	$M = \leq 1.5x10^6$ (rolling geometric average over a 2-month period, with at least 2 samples per month)	If standards exceeded, inform the competent authority and take steps to correct the situation.
Raw milk from species other than cows at milk production holdings, if intended for manufacture	Plate count (30° C)	$M = \leq 5x10^5$ (rolling geometric average over a 2-month period, with at least 2 samples per month)	If standards exceeded, inform the competent authority and take steps to correct the situation.
Raw cows' milk used to prepare dairy products immediately before processing	Plate count (30° C)	$M = < 3x10^5$	If standards exceeded, inform the competent authority and take steps to correct the situation.
Processed cows' milk used to prepare dairy products	Plate count (30° C)	$M = < 10^5$	If standards exceeded, inform the competent authority and take steps to correct the situation.

6.7. OTHER RELATED NON-MICROBIOLOGICAL STATUTORY HYGIENE CRITERIA

LEGISLATION	FOOD CATEGORY	MICROORGANISM	LEVEL (per g or ml, unless otherwise specified)	STATUS/ACTION
Commission Regulation (EC) No. 853/2004 as amended of the European parliament and of the Council laying down specific hygiene rules for food of animal origin	Raw cows' milk and colostrum at milk production holdings	Somatic cell count	M = ≤ $4x10^5$ (rolling geometric average over a 3-month period, with at least 1 sample per month)	If standards exceeded, inform the competent authority and take steps to correct the situation.
Commission Regulation (EC) No. 2074/2005as amended, as regards implementing measures for certain products of animal origin intended for human consumption and repealing certain implementing measures.	Heat-treated milk	Alkaline phosphatase activity	M = ≤ 350 milliunits (mU) of enzyme activity/l	If standards exceeded, withdraw from the food chain, inform the competent authority and take steps to correct the situation.

APPENDIX I: SPECIFIC ORGANISMS MICROORGANISMS AND THE PRINCIPLES OF TEST METHODS

AEROBIC COLONY COUNT

Aerobic colony count (ACC; synonyms include aerobic colony count, standard plate count, total viable count (TVC), or aerobic mesophilic colony (or plate) count) is one of the most common methods used to indicate the total microbiological load of the product. It can indicate the quality of the product and/or whether a product has been contaminated, temperature abused or spoilt. The type and range of microorganisms making up an ACC will depend on the growth conditions but could include bacterial species belonging to the genera of Bacillus, *Pseudomonas*, Enterobacteriaceae as well as yeasts.

Samples are plated onto non-selective, nutrient rich media and incubated aerobically, usually at 30°C but other temperatures are also common. The temperature and length of incubation should be defined. The most common incubation regime is 30°C for 48 or 72 hours for mesophilic microorganisms, but other temperatures and times may be more appropriate e.g. 55°C for 48-72 hours for thermophilic microorganisms or 20-25°C for up to 5 days for psychrotrophic microorganisms. ISO recommends 6.5°C for 10 days for psychrotrophic microorganisms (ISO, 2001). Different temperatures of incubation will lead to isolation of different flora; many microorganisms that develop during chilled storage for example will not grow well above 30°C, and so a lower temperature of incubation is preferable i.e. 20-25°C. For enumeration of aerobic colony count, ISO 4833-1 should be followed.

Consideration should be given to the method used for performing an aerobic colony count. Pour plate methods are frequently specified, but microorganisms commonly found in foods may be injured by the temperature of the molten agar used especially if they are already stressed. In these circumstances, surface inoculation methods may be preferable.

Anaerobic colony counts can be used for foods that provide anaerobic conditions for microbial growth e.g. processed cheese, vacuum packed products, and canned foods. Consideration should also be given to foods in which conditions are, or may become, anaerobic (reduced redox potential) such as deep in meat tissues.

AEROMONAS HYDROPHILA AND RELATED SPECIES

The role of *Aeromonas* in causing food poisoning has not been clearly established, but some strains of certain species are thought to can cause diarrhoea. To date, a clonally defined outbreak has yet to be confirmed and no animal model exists with which to reproduce the disease.

In the first infectious intestinal disease study (IID, 2000), *Aeromonas* were found to be as common in controls as cases of disease and in the second study (IID, 2012), *Aeromonas* spp. were not included because if their doubtful pathogenicity and clinical significance. They are aquatic microorganisms but have been isolated from many foods. *Aeromonas* spp. grow at refrigeration temperatures. Enumeration techniques on solid media and enrichment methods to establish presence or absence are both appropriate. This group of microorganisms is not normally included in microbiological specifications.

ALICYCLOBACILLUS

Alicyclobacillus is an acid-tolerant, Gram positive spore-former. The most common of the 18 species is *A. acidoterrestris.* This genus of bacteria is able to grow at low pH values (2.5-3.5), tolerate temperatures of 50-70°C and produce heat-resistant spores. These bacteria are able to grow at low pH values (minimum pH for growth around 2.5), tolerate temperatures of 50-70°C and produce heat resistant spores. The minimum temperature for growth is circa 25°C, so storage of fruit beverages below this temperature will not permit growth and subsequent spoilage. Often a 'medicinal' or 'antiseptic' off-odour (caused by production of guaiacol) of fruit juices denotes *Alicyclobacillus* spoilage.

This bacterium is found in soil, extreme environments (e.g. hot springs) and acidified fruit juices. There are generally no limits for numbers of the microorganism in foods, but it is recognised that the presence of low numbers in processed fruit juices stored at elevated temperatures can result in taint formation. To mitigate against problems arising it is important to use adequate heat processing and rapid cooling in processing. Consideration can be given to the use of the preservatives sorbic or benzoic acid to prevent spore germination.

BACILLUS CEREUS AND OTHER *BACILLUS* SPECIES

The members of the Bacillaceae produce spores as well as vegetative cells. The spores can survive adverse conditions such as drying and certain heat treatments including pasteurisation. These microorganisms are common in the environment and in many foods. Published reports of *B. cereus* food poisoning show that high levels (>10^5 per gram) in food clearly cause illness and in that in rare cases, levels of 10^3 spores/g have caused illness (EFSA, 2004). *Bacillus* species can also cause spoilage.

The most common cause of *Bacillus* food poisoning is that due to *B. cereus*; two forms of food poisoning have been reported, due to different types of toxin. One type is emetic, causing nausea and vomiting due to ingestion of toxic cereulide compounds preformed in the food, whilst the second is diarrhoeal due to the effect of an enterotoxin produced after ingestion of large numbers of the microorganism which produce enterotoxin in the gut and/or food although enterotoxin is not considered to survive passage through the stomach due to effects of digestive enzymes and low pH.

The emetic syndrome is particularly associated with starchy products such as rice and pasta dishes, whereas a wider range of foods have been implicated with the diarrhoeal syndrome, including meat products, soups, vegetables and sauces. The emetic toxin of *B. cereus* is particularly heat resistant (heat stable to 126°C for 90 minutes) whereas the diarrhoeagenic toxin is inactivated by exposure to 56°C for 5 minutes (Jensen & Moir, 2003). Pasteurization inactivates diarrhoeagenic toxins (EFSA, 2016).

Food poisoning caused by the *B. subtilis* group (*B. subtilis, B. licheniformis, B. pumilis* and *B. amyloliquifaciens*) occurs less frequently than that caused by *B. cereus*, with meat and pastry products and rice dishes implicated (Kramer and Gilbert 1989). In addition, some strains of *B. thuringiensis* have been shown to produce enterotoxin. As members of this group are used in insecticide preparations, non-toxigenic strains must be carefully selected. Difficulty in investigating food poisoning incidents can arise due to the high degree of similarity between *B. thuringiensis* and *B. cereus*, and their appearance on selective media.

Recovery and enumeration of *Bacillus* species can be accomplished on many media, but media designed specifically for enumeration of presumptive *B. cereus* are usually used. These media may allow resuscitation, growth from spores and identification of presumptive *B. cereus* if present. Atypical strains of *B. cereus* can occur and could lead to a false negative result. For example, some strains of *B. cereus* produce little or no lecithinase. Colonies of these strains will not be surrounded by the precipitation zone on polymixin pyruvate egg yolk mannitol agar (PEMBA) or mannitol phenol red egg yolk polymixin agar (MEYP) plates.

Besides these traditional diagnostic media, a range of chromogenic media are now available for the detection and isolation of *B. cereus* and *Bacillus thuringiensis* from foods. In some circumstances mesophilic or thermophilic spore counts may be required; methods incorporate heat treatment of the sample prior to spore enumeration at appropriate temperatures. Biochemical tests (including ammonium salt sugars (glucose, arabinose, mannitol and xylose) and a nitrate reduction test) can be used to distinguish the most common food poisoning *Bacillus* species (Roberts and Greenwood 2003).

Rapid method kits are available for detecting *B. cereus* diarrhoeal toxin (Campden BRI, 2012) and also for emetic toxin-producing strains. Methods to detect emetic toxin include cell culture assays and liquid chromatography/mass spectrometry. There is a published standard method in ISO 18465:2017 (ISO, 2017b) enabling quantitative analysis of the emetic toxin cereulide using high performance liquid chromatography (HPLC) or ultra-performance liquid chromatography (UHPLC) connected to a tandem mass spectrometer (LC-MS/MS).

CAMPYLOBACTER SPECIES

Campylobacter jejuni and other thermotolerant *Campylobacter*s are the most common cause of bacterial diarrhoea but with the exception of poultry, the sources and routes of transmission have still not been fully determined (EFSA, 2005).

Campylobacter are found in the intestinal tracts of many animals, especially poultry and other birds, and can therefore contaminate water sources and raw milk as well as meat and poultry. These microorganisms have fastidious temperature and atmospheric requirements for growth. They do not grow in foods under normal storage conditions. As the infectious dose is low (a few hundreds of cells) (Blackburn & McClure, 2009), the presence of any in ready to eat foods is significant. Isolation of the microorganism from food and environmental sources is, therefore, carried out using enrichment methods and selective agars. Regulation (EC) 2073/2005 testing for *Campylobacter* in the carcasses of broilers is a requirement of the process hygiene criteria.

CLOSTRIDIUM PERFRINGENS AND SULPHITE REDUCING CLOSTRIDIA

Clostridium perfringens is a spore-forming anaerobe commonly found in mammalian faeces and soil. The spores persist in the environment and often contaminate raw food materials. Spores may survive cooking and subsequent germination and rapid growth of vegetative cells may occur if the food is not chilled effectively. Outbreaks caused by this organism are frequently associated with foods held warm, as it has an optimum growth temperature around 43-45°C. If large numbers of the microorganism are eaten in a food, they can sporulate in the gut and produce an enterotoxin which causes diarrhoea.

Isolation of *C. perfringens* on solid media requires anaerobic conditions of incubation, and usually the use of selective agents, followed by confirmation of suspect colonies. Vegetative cells of *C. perfringens* do not survive well in foods at low temperatures, and spore counts may be more appropriate on such foods.

The sulphite reducing clostridia (SRCs) include the pathogens *C. perfringens* and *C. botulinum* (the causative microorganism of botulism). The SRCs may be used as an indicator of plant hygiene and for the presence of other clostridia that may cause food poisoning or food spoilage. *C. botulinum* is sub-divided by heat resistance of spores and temperature limits for growth, into proteolytic and non-proteolytic groups. Proteolytic species produce very heat resistant spores and vegetative cells can only grow down to 10°C whereas the spores of non-proteolytic species are easily destroyed at 90°C for 10 minutes, but the vegetative cells can grow down to 3°C and as such are associated more with chilled foods.

There are no routine laboratory tests specific enough for confirming the presence of *C. botulinum* and specialist testing is required for its isolation and detection of its toxins. It is also not known at what point during growth the toxins may be produced, so evidence of growth is a cause for concern. Control strategies are thus aimed at preventing germination of spores and subsequent growth of the organism and therefore routine testing is generally not useful.

CRONOBACTER SPECIES (FORMERLY *ENTEROBACTER SAKAZAKII*)

Cronobacter species (including formerly *Enterobacter sakazakii*) now comprise a new genus with species, including *sakazakii* that have been linked to a rare cause of invasive infection in new-born babies. This has included several outbreaks of neonatal meningitis and necrotising enterocolitis in the past. Infection by *Cronobacter* spp. has a high mortality rate estimated between 40-80% in vulnerable infants. The few outbreaks of infection that have occurred have implicated contaminated powdered formula milk as the source of the infection. These bacteria do not survive the pasteurisation process of powdered formula milk production. However, this milk may have become contaminated following pasteurisation, during the addition of micronutrients or during reconstitution.

The natural environment of *Cronobacter* spp. is not known though it is suspected to be a widespread microorganism. They have been isolated from several environments including hospitals, households, cereal factories and milk powder factories. They are also noted for their ability to survive for extended periods under dry environmental conditions.

The EC Regulation on microbiological criteria for foodstuffs ((EC) No. 2073/2005) provides that *Cronobacter* spp. should be absent from dried infant formula and dried dietary foods for special medical purposes intended for infants below six months of age. Methods for the detection of *Cronobacter* spp., including a published standard method in ISO 22964:2017 (ISO, 2017d) which involves selective enrichment followed by subculture to selective agar. Rapid methods (e.g. PCR-based assays) have also been developed for detection of *Cronobacter* spp.

ENTEROBACTERIACEAE AND COLIFORMS

The family Enterobacteriaceae is a large taxonomic group that includes several pathogens (e.g. *Salmonella* spp., *Shigella* spp., *Yersinia enterocolitica*, and pathogenic *E. coli*), many harmless commensal species that inhabit the gut, and a few environmental microorganisms and plant pathogens, e.g. species of *Escherichia, Citrobacter, Enterobacter, Proteus*.

Enterobacteriaceae are associated with the intestinal tracts of animals (including humans), plants, soil, water and other environmental sources. Hence, they can act as general indicators of hygiene in food manufacturing environments. Enterobacteriaceae (including faecal indicators such as generic *E. coli* that is closely associated with the intestines and faeces) are useful indicators of hygiene and of post-processing contamination of heat processed foods, as they should be inactivated by the heat process. However, in minimally processed foods such as salad leaves and other raw or minimally heated products, levels may be high. These types of products should be routinely tested for faecal indicators and action taken when upward trends are observed. As low levels may be tolerated, most methods for Enterobacteriaceae are based on enumeration using direct plating on media containing bile salts and glucose. However, if low levels are not tolerated, or if the food is likely to contain sub-lethally injured cells, a detection method, or a method including a resuscitation step, should be selected. These include MPN-based method which can provide quantitative results. The colony count method for Enterobacteriaceae is specified in ISO 21528-2 (ISO,

Coliforms are a subgroup of Enterobacteriaceae; while this is not a taxonomic group it is often defined by the methods used for detection and isolation. These depend on the organisms' ability to ferment lactose rapidly or the presence of β-galactosidase activity. The coliform group contains species belonging to the genera *Enterobacter, Citrobacter, Escherichia* and *Klebsiella* but others may also be included.

Thermotolerant coliforms, a more restricted group of microorganisms, are those coliforms that can multiply at 44°C. Most faecal coliforms are also able to multiply at 44°C and in practice the terms 'faecal' and 'thermotolerant' are used synonymously. This group includes *E. coli* types I and II and occasional strains of *Klebsiella* and *Enterobacter*. Psychrotrophic strains of Enterobacteriaceae, including some coliform strains may be of interest to manufacturers of chilled foods, notably dairy products.

Figure 2. The relationship between Enterobacteriaceae, coliforms, faecal coliforms, *Salmonella*, *E. coli* and STEC.

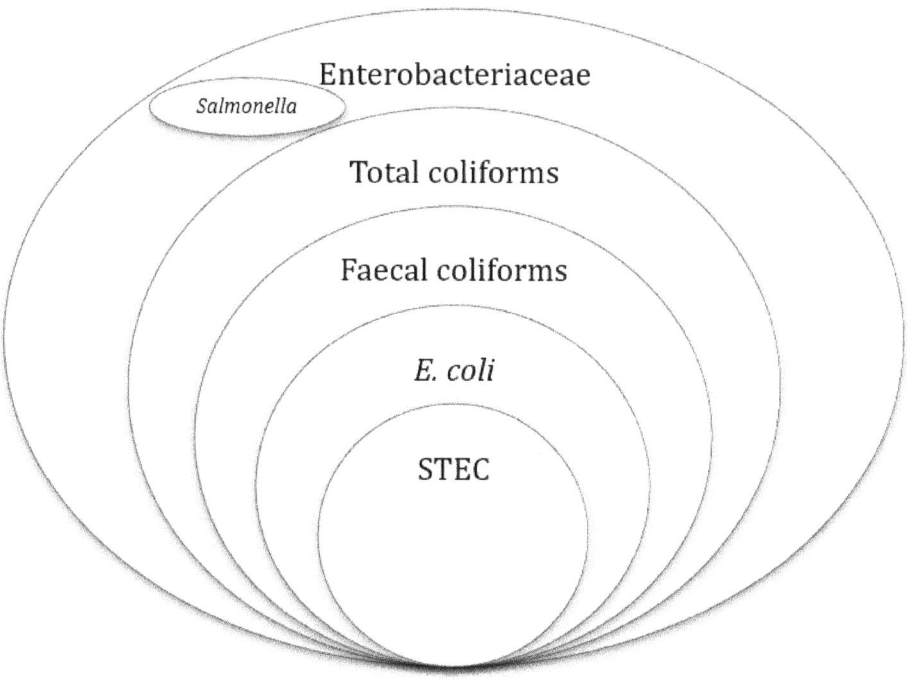

Enterococcus species

The enterococci, which include species formerly termed faecal streptococci, survive and grow well in factory environments, especially dry environments. They are more resistant to heat, freezing and adverse conditions than the Enterobacteriaceae and therefore, they are used as alternative hygiene indicators, particularly in dried products and frozen foods.

Methods for the detection and enumeration of enterococci usually involve isolation on selective agar.

Escherichia coli and STEC

Generic or commensal *E. coli* form part of the Enterobacteriaceae family and are mostly harmless - causing no illness in the host. The presence of *E. coli* in food, however, is generally undesirable because this species is commonly associated with the gastrointestinal tract and faeces so their presence indicates poor hygienic conditions and potential faecal contamination, which may indicate the presence of other enteric pathogens such as *Salmonella*. Moreover, certain *E. coli* strains are pathogenic and may cause gastroenteritis as well as a range of other foodborne diseases. Absence of *E. coli* is normally required in processed foods, and so a detection method or MPN method is required to detect low levels. Further specificity can be introduced by using a media or a method employing fluorogenic or chromogenic substrates diagnostic for β-glucuronidase activity; an enzyme possessed by most strains of *E. coli* and relatively uncommon in other bacteria. Direct enumeration methods for coliforms and *E. coli* may be more suitable when contamination might be expected e.g. raw meats. Caution should be used in comparing results obtained by different methods and it is important to clearly define the target microorganism(s) and method used.

There are several *E. coli* strains that are pathogenic, and these are generally assigned to a particular pathotype. One group of pathogenic *E. coli*, responsible for mild to serious life threatening disease is the Shiga toxin-producing *E. coli* (STEC) The term STEC has been used interchangeably with VTEC (Verocytotoxin-producing *E. coli* or verocytotoxigenic *E. coli*), Within the STEC is a sub-group termed the Enterhaemorrhagic *E. coli* (EHEC) which are defined by the carriage of additional virulence associated genes and more serious clinical conditions.

Infection with STEC can cause symptoms ranging from mild diarrhoea to severe bloody diarrhoea (haemorrhagic colitis), sometimes progressing to haemolytic uraemic syndrome (HUS) and kidney failure. The infective dose of *E. coli* O157:H7, one of the most common serovars of STEC associated with foodborne disease, is thought to be very low (e.g. 10s of cells) (Bell & Kyriakides, 2009). The presence of any STEC in a ready to eat food is considered a potential hazard. Other pathotypes of *E. coli* include enterotoxigenic *E. coli* (ETEC), enteroinvasive *E. coli* (EIEC), enteroaggregative *E. coli* (EAEC), and enteropathogenic *E. coli* (EPEC) and diffusely adherent *E. coli* (DAEC), can cause a range of diseases depending on the pathotype.

A precise definition of what makes an STEC pathogenic has yet to be defined although it is thought that the virulence factors an STEC contains are likely to be the most important predictive factors, and these should be considered alongside any epidemiological evidence, rather than on the serogroup alone. This is also complicated by other factors, for example a large outbreak in Germany in 2011 was caused by serotype O104:H4, which was first thought to be an EHEC strain but was later shown to be an EAEC that had acquired the *stx* genes, highlighting the complexity of the situation and importance of horizontal gene transfer in the emergence of new pathogens.

Outbreaks of foodborne illness associated with STEC have been linked to several commodities, including undercooked beef, fermented meats, vegetables such as lettuce and spinach, dairy products including unpasteurised milk and cheeses, and unpasteurised juices. In 2011 the *E. coli* O104:H4 caused one of the most significant outbreaks of STEC infection recorded in Europe, including 855 cases of HUS, and 53 deaths and was associated with the consumption of sprouted fenugreek seeds.

Methods for the enumeration of generic *E. coli* and detection of *E. coli* O157 are readily available, however, the other non-O157 STEC strains require more specialist techniques. It should be noted that the elevated incubation temperatures used for the recovery of faecal coliforms and *E. coli* are unsuitable for the growth of some pathogenic *E. coli*. Many strains of *E. coli* O157:H7, for example, do not grow well at 44°C and most strains of this serotype do not show β-glucuronidase activity.

Lactic acid bacteria

This group includes bacteria capable of producing lactic acid from fermentable substrates. Some strains grow poorly in aerobic conditions. Genera commonly found in/on food include *Lactobacillus, Pediococcus, Leuconostoc, Lactococcus* and *Streptococcus*. They are ubiquitous in the environment and are notable for their ability to grow at low pH levels.

Starter cultures for fermented foods such as cheese and salami commonly include this group of microorganisms, and so their presence in high numbers should be expected. However, in some foods such as raw pastry, they may cause spoilage through souring. In other foods (e.g. vacuum-packaged cooked ham), heterofermentative lactic acid bacteria can produce carbon dioxide gas as well as lactic acid to give rise to spoilage. Direct enumeration methods on solid media are usually used to detect these microorganisms.

Listeria species

Listeriosis, caused by *L. monocytogenes*, is a foodborne infection of great public health concern due to its clinical severity and high mortality rate, particularly in pregnant women, new-borns, the elderly and the immunocompromised. *L. monocytogenes* is widely distributed in the environment and the microorganism is therefore present in a wide variety of raw food materials. *L. monocytogenes* is resistant to diverse environmental conditions, able to grow at refrigeration temperatures, and survive in food for prolonged periods under adverse conditions. Its ability to form biofilms and colonise food processing environments is also well recognised.

The European Commission (EC) Scientific Committee on Veterinary Measures relating to Public Health (SCVPH) considered the risk of listeriosis to be low from foods containing levels of 10^2 cfu/g and below. As a result, the EC Regulation on microbiological criteria for foodstuffs ((EC) No. 2073/2005) specifies that *L. monocytogenes* should be below 10^2 cfu/g during the shelf-life of ready-to-eat foods, and that processing areas and equipment used in the manufacture of ready-to-eat foods must also be monitored for *L. monocytogenes*. While *L. monocytogenes* is the species of concern, foods are sometimes examined for *Listeria* spp. as an indicator of hygiene and the likely risk of *L. monocytogenes* being present. Some *Listeria* species are notably unable to grow at low refrigeration temperatures and as such are not as useful as indicators. Testing for *Listeria* only requires identification to exclude *L. monocytogenes*. Identification of other species is of little use.

Detection (presence/absence) tests for *L. monocytogenes* and other *Listeria* species involve selective enrichment followed by subculture to selective agar. Enumeration of *L. monocytogenes* and other *Listeria* spp. is usually performed on solid selective media.

PROTOZOAN PARASITES

The most important waterborne or foodborne protozoan parasites that infect humans are *Cryptosporidium* spp. (particularly *C. parvum, C. hominis* and *C. meleagridis*), *Cyclospora cayetanensis* and *Giardia lamblia, Sarcocystis* spp. and *Toxoplasma gondii.*

Outbreaks of protozoal infection in humans are most often associated with contaminated water or food handlers and have been linked to raw fruits and vegetables. The presence of these protozoa on raw fruits and vegetables is likely to be due to contact with animal or human faeces, sewage, water containing untreated sewage, and sludge. Strict codes of hygiene control in growing and harvesting areas are therefore most important. For example, the Biosolids Nutrient Management Matrix (ADAS, 2001a; ADAS, 2001b) and the UK Chilled Food Association (CFA 2002) provide guidelines for produce suppliers to help minimise microbial food safety hazards.

C. hominis and *C. cayetanensis* appear to be specific to humans, hence all infections will involve direct contact with another infected person or, food, water or beverages that have become contaminated with human sewage. Transmission of *Giardia* and *Cryptosporidium* spp. is mainly by ingestion of contaminated water but similarly, ingestion of contaminated food and person-to-person transmission can also occur. For the detection of these parasites in water, methods involving concentration by filtration of large volumes of water are required. For surface water environmental monitoring, grab samples of 10 - 20 litres might be suitable, and for water supply and distribution system monitoring, volumes up to 100, or even 1000 litres (employing on-site concentration) may be appropriate. In the UK, The Water Supply (Water Quality) (Amendment) Regulations 1999 regulations require that the average number of cryptosporidial oocysts is less than one per 10 litres (L) of potable water. To verify compliance with this requirement water companies must ensure that the water leaving the relevant treatment works is continuously sampled and analysed daily for cryptosporidial oocysts. Methods for routine detection in food are poorly developed.

Transmission of *Sarcocystis* results from ingestion of raw or undercooked meat containing mature cysts; these may be detected by the naked eye or microscopic examination of histological sections. By comparison, the main host of *T. gondii* is the cat which excretes oocysts into the environment. These can lead to infection in food animals. Ingestion of raw or undercooked meat may cause infection as may direct contact with infected cats and consumption of contaminated water. These parasites are rarely sought in food and water unless they are implicated in illness.

PSEUDOMONAS SPECIES

Pseudomonas species are ubiquitous in the environment and water, and some species are psychrotrophic. *Pseudomonas aeruginosa* is of clinical significance as an opportunistic pathogen but has rarely been implicated in gastroenteric infection. Other pseudomonads are significant in food spoilage, particularly in chilled food. Levels higher than 10^5-10^7 cfu/g or ml of food may result in off flavours, off odours and visual defects. Direct enumeration methods on selective solid media are used.

SALMONELLA SPECIES

There are two species of *Salmonella* (*enterica* and *bongori*) and over 2500 serotypes. They belong to the family Enterobacteriaceae, are microorganisms of faecal origin, and are all regarded as pathogenic to humans. *Salmonella* spp. are facultatively anaerobic mesophiles. They are not particularly heat resistant and are normally destroyed by pasteurisation. *Salmonella* may survive for long periods in e.g. dried faeces or food materials.

Salmonella spp. cause significant numbers of food poisoning cases each year. Outbreaks have been associated with a variety of foods and are commonly associated with poultry and eggs. Classical methods for detection of *Salmonella* spp. involve testing a defined amount of food, usually 25g. Large quantities should be examined for food products for which increased assurance is required e.g. some dried foods, or if the food is destined for a susceptible population e.g. baby foods (see Section 3).

Methods must be able to detect low levels of *Salmonella* spp. in the presence of high levels of background flora. Traditional procedures used to achieve this are lengthy, and typically involve a pre-enrichment resuscitation stage in a non-selective medium, enrichment in selective broth media, and subculture to solid selective media. The ISO method for detection, enumeration and serotyping of *Salmonella* spp. is ISO 6579-1 (ISO 2017e). Atypical (and rare) lactose-fermenting *Salmonella* spp. and hydrogen sulphide negative strains, may not be detected by many common test methods. No single medium or method is optimal for growth of all serotypes of *Salmonella*; two formulations each of enrichment and solid selective media are often used in food examinations. Chromogenic media now available for *Salmonella* detection and isolation can improve isolation of atypical *Salmonella*s as these media do not rely on the biochemical reactions employed by traditional selective diagnostic *Salmonella* media. Biochemical and serological confirmation of suspect colonies are required. More rapid methods have been, and continue to be, developed for screening out negative samples and automating test procedures. Any such alternative methods should be assessed by each laboratory for the types of foods that they are required to examine.

SHIGELLA SPECIES

Members of the Enterobacteriaceae, *Shigella* spp. are similar in their general growth and survival characteristics to *Salmonella* spp., although less persistent in the environment. Shigellosis, also called bacillary dysentery, is caused by four species: *Shigella dysenteriae, Sh. flexneri, Sh. boydii* and *Sh. sonnei*.

Sh. sonnei usually causes a mild illness. *Sh. boydii* and *Sh. dysenteriae*, and most *Sh. flexneri* infections present clinically as dysentery. They are highly infectious enteric pathogens, with an infective dose as low as 10 cells so enrichment methods of detection are used for isolation. The principal route of transmission is faecal-oral from cases with diarrhoea. Food may be contaminated by infected food-handling personnel or by contaminated irrigation water. Foodborne illness due to *Shigella* is relatively uncommon in the UK and *Shigella* species are rarely included in microbiological criteria.

STAPHYLOCOCCUS AUREUS AND COAGULASE-POSITIVE STAPHYLOCOCCI

Coagulase-positive staphylococci include *Staphylococcus aureus* (*Staph. aureus*), which may be associated with the skin, nose and throat of healthy individuals as well as being the cause of boils and many skin and wound infections. The microorganism will also occur naturally in poultry and other raw meats as a frequent component of the skin microflora, and also in raw milk where levels may sometimes be elevated as a result of *Staph. aureus* mastitis in the producing herd. Given suitable conditions it is capable of growth in foods, and some strains may produce a heat-stable enterotoxin that can survive the heat processes applied to low-acid canned foods, as can some other coagulase- positive and negative staphylococci such as strains of *Staph. intermedius* and *Staph. hyicus.*

Testing for coagulase-positive staphylococci is used as an indicator of general and personal hygiene and adherence to good food handling procedures. Their presence, particularly in high numbers ($>10^4$ per gram) in foods, may indicate the potential presence of enterotoxin and a consequent risk of food poisoning. If poor temperature control allowing growth to high numbers has occurred prior to heat treatment, then the enterotoxin may remain despite the absence of viable cells. There have been outbreaks caused by milk products where growth and enterotoxin production occurred in the raw milk and the enterotoxin, but not the microorganism, survived the subsequent manufacturing process, e.g. pasteurisation.

The European Commission Regulation (EC) 2073/2005 therefore lays down food safety criteria and process hygiene criteria for staphylococcal enterotoxins and coagulase-positive staphylococci respectively, in cheese, milk powder and whey powder. The group is normally detected by direct selective enumeration techniques with confirmation via coagulase tests, and *Staph. aureus* via selection enumeration. The ISO method for detection of staphylococcal enterotoxins is ISO 19020 (ISO, 2017f) which comprises an extraction step followed by immunoassay, using commercially available kits. Additionally, PCR methods are available that identify the genes that encode for the enterotoxins A – E, G – R, U, and V.

VIBRIO PARAHAEMOLYTICUS AND OTHER PATHOGENIC *VIBRIO* SPECIES

Vibrio parahaemolyticus and other *Vibrio* spp. are found in warm, shallow coastal waters and are contaminants of fish and shellfish. Foodborne infections caused by *V. parahaemolyticus* are therefore almost exclusively associated with the consumption of raw, improperly cooked or cooked and re-contaminated fish and shellfish. Levels above 10^2 cfu/g in food are regarded as unacceptable.

There have been several outbreaks associated with *Vibrio* species, which are normally associated with a diarrhoeal illness, nausea and vomiting, some outbreaks have however resulted in deaths, particularly in vulnerable groups. Illness and outbreaks are primarily associated with the consumption of raw or inadequately cooked oysters. Enrichment methods are normally used for their isolation although enumeration can be useful. Improvements to detection and isolation have been achieved using molecular techniques such as PCR. There are now molecular-based methods for the improved detection of *V. parahaemolyticus*, *V. cholerae* and *V. vulnificus* e.g. ISO 21872-1:2017 (ISO, 2017c).

V. cholerae is the causative microorganism of cholera and is usually transmitted via contaminated water and the faecal-oral route. The microorganism is not usually included in routine tests for foods except in response to specific public health concerns. Several other species of *Vibrio* are of pathogenic significance e.g. *V. vulnificus*, and also appear to be associated with consumption or handling of contaminated seafood. The presence of

Vibrio species in cooked seafood indicates either inadequate heat treatment or post-processing contamination.

VIRUSES

Foodborne viral illness is caused mainly by enteric viruses that cause gastroenteritis. Common examples include norovirus, hepatitis A and E, and rotavirus. Viruses cannot grow in foods but can be carried by foods if they become contaminated. The infectious dose of enteric viruses is very low (~1-10 particles in the case of norovirus) (Duizer & Koopmans, 2009). Primary contamination of the food at source is caused mainly by polluted water, contaminated irrigation water and contaminated handlers hands. Water which is used for pesticide preparation may also be a source of contamination if it is used untreated. Consumption of raw or lightly processed molluscan shellfish harvested from sewage polluted waters has been implicated in many outbreaks of gastroenteritis; contaminated soft fruits, vegetables and salad items which do not receive a heat process may also cause illness if they have become contaminated with enteric viruses.

Many viral foodborne illnesses result from secondary contamination of food by infected food handlers during preparation of items which do not receive further heat treatment. There is widespread presence of hepatitis E virus in pig herds, both in the UK and elsewhere in Europe, and there has been a large increase in the number of indigenous hepatitis E cases in recent years. Hepatitis E illness can develop when pork products originating from hepatitis E infected pigs is eaten raw or minimally processed.

Detection of viruses in foods (norovirus and hepatitis A virus) can be achieved using standard methods (ISO, 2017a; ISO 2019). These are based on the detection of the genetic material from the viruses.

Currently there are no microbiological specifications for viruses in foods, however there is a zero-tolerance stance being taken by most authorities when detected during routine monitoring.

YEASTS AND MOULDS

Yeasts are a common cause of food spoilage, particularly of acid foods such as fruit and fruit juices, and foods of reduced water activity (a_w) such as confectionery. They have not been implicated in food poisoning. Moulds are also a significant cause of food spoilage, and some yeast and mould strains can grow at very low a_w values e.g. 0.6-0.7, causing spoilage of otherwise microbiologically stable commodities such as bakery goods. Some can produce mycotoxins which can cause serious chronic illness if consumed, although acute illness can also occur.

Moulds can rarely be successfully enumerated when growing in foods. A focus of mould growth may consist either of hyphae or of hyphae producing spores. 'Counts' obtained from hyphae will vary depending on the degree of fragmentation caused by macerating the sample and can range from 10s to many hundreds per gram. If spores are present, the 'count' may rise to millions per gram. In these circumstances, 'counts' are usually meaningless. A more meaningful test is to incubate the sample and assess for growth to a visible level within the shelf life of that product.

Selective agars are commonly used for the detection of both yeasts and moulds. The brewing industry use specific agars for detecting 'spoilage yeasts'. Surface inoculation methods of enumeration are preferable to pour plate methods because of the need to maximise aerobic conditions for mould growth and to enhance colony morphology for identification purposes. Incubation is normally lengthy and can take 5 days (or considerably longer for xerophiles) to obtain a result.

Standard enumeration tests are inappropriate when the significant microorganism is present as a very low proportion of the contaminant flora e.g. heat resistant moulds on fruit; or when significant contamination is below the detection limit of the method e.g. a few yeasts in a 200 litre drum of fruit juice concentrate. In such circumstances and if deemed necessary, enrichment or concentration procedures may be employed for detecting relevant microorganisms.

YERSINIA ENTEROCOLITICA AND RELATED SPECIES

Yersinia species are also members of the Enterobacteriaceae and contain 18 species including *Y. pseudotuberculosis* and *Y. enterocolitica*, which are pathogenic for humans and animals. They are widely distributed in animals of all types and are found in many environments. *Y. enterocolitica*, the species most often associated with foodborne yersiniosis, is psychrotrophic, and is mainly associated with pigs, although food vehicles other than pork are frequently associated with outbreaks. Strains of *Y. enterocolitica* can cause gastroenteritis and many other symptoms. The infectious dose is not known however, the presence of pathogenic serotypes should be regarded as potentially hazardous.

Other species of *Yersinia* and their serotypes have been found in a wide variety of foods, and their presence in heat-processed foods indicates post-processing contamination. Isolation is usually accomplished using enrichment methods. Identification of *Yersinia* spp. is still based mainly on biochemical tests and serotyping, but molecular methods are increasingly also being used. These microorganisms are not normally included in microbiological criteria.

REFERENCES

ADAS (2001a). The Safe Sludge Matrix. Guidelines for the Application of Sewage Sludge to Agricultural Land. ADAS, Water UK and BRC. 3rd Edition. Accessed June 2019. Available at: http://adlib.everysite.co.uk/resources/000/094/727/SSMatrix.pdf

ADAS (2001b). The Safe Sludge Matrix. Guidelines for the Application of Sewage Sludge to Industrial Crops. ADAS, Water UK and BRC. 3rd Edition. Accessed June 2019. Available at: http://adlib.everysite.co.uk/resources/000/094/763/ASSIC.pdf

ACMSF (1992). Advisory Committee on the Microbiological Safety Of Food. Report on Vacuum Packaging and Associated Processes. London, HMSO.

AOAC (1995). The Association of Official Analytical Chemists (AOAC). Official Methods of Analysis of the AOAC, 16th Edition, AOAC: Arlington, Virginia, USA.

ARCHER, D.L. (2004) Freezing: an underutilized food safety technology? Int. J. Food Microbiol. 2004, 90, 127-138.

BELL, C. & KYRIAKIDES, A. (2009) Pathogenic *Escherichia coli.* Chp 17 p. 581-626 In 'Foodborne Pathogens', Eds. Clive de W. Blackburn & Peter J. McClure, 2nd Edition, ISBN: 978-1-84569-362-6 Woodhead Publishing, UK.

BELL, C., NEAVES, P. & WILLIAMS, A. P. (2005) Food Microbiology and Laboratory Practice. Oxford, UK, Blackwell Publishing.

BLACKBURN, C.W. de & McCLURE, P.J. (2009) Campylobacter and Arcobacter. Chp. 20 p.718-762 In 'Foodborne Pathogens', Eds. Clive de W. Blackburn & Peter J. McClure, 2nd Edition, ISBN: 978-1-84569-362-6 Woodhead Publishing, UK.

BRACKETT, R.E., OCASIO, W., WATERS, K., BARACH, J., & WAN, J. (2014) Validation and Verification: A Practical, Industry-driven Framework Developed to Support the Requirements of the Food Safety Modernization Act (FSMA) of 2011. Food Protection Trends November/December 410-425.

BRITISH STANDARDS INSTITUTION (BSI) (2015). BS EN ISO 7218:2007 +A1:2013 Microbiology of food and animal feeding stuffs. General requirements and guidance for microbiological examinations. London, England: BSI. 2015

CAMPDEN BRI (2004) Guideline 46: Evaluation of product shelf-life for chilled food. ISBN:0-905942-65-5 Campden BRI, UK.

CAMPDEN BRI (2006) Guideline 51: Pasteurisation: A Food Industry Practical Guide. 2nd Edition ISBN:978-0-905942-896 Campden BRI, UK.

CAMPDEN BRI (2007) Guideline 43: Manual of Microbiological Methods for the Food and Drinks Industry. 5th Edition ISBN:978-0-905942-93-3 Campden BRI, UK.

CAMPDEN BRI (2012) Review 1: Catalogue of rapid microbiology methods. 7th Edition ISBN:978-0-907503-72-9 Campden BRI, UK.

CAMPDEN BRI (2013) Guideline 9: A code of practice for microbiology laboratories handling food, drink and associated samples. 4th Edition ISBN:978-0-907503-74-3 Campden BRI, UK.

CAMPDEN BRI (2015) Guideline 74: Validation and optimisation of thermal processing systems: cookers, pasteurisers and sous vide systems. ISBN:978-0-907503-80-4 Campden BRI, UK.

CAMPDEN BRI (2017) Guideline 52: Establishment and use of microbiological criteria (standards, specifications and guidelines) for foods. 2nd Edition ISBN:978-0907503-90-3 Campden BRI, UK.

CHILLED FOOD ASSOCIATION (CFA). (2002). Microbiological Guidance for Produce Suppliers to Chilled Food Manufacturers (Micro Guidance for Growers). CFA Ltd, London.

CODEX (1981): Report of 17th Session of the Codex Committee on Food Hygiene, Washington, DC. 17-21 November 1980. ALINORM 81/13. Appendix II, General Principles for the Establishment and Application of Microbiological Criteria for Foods. Codex Alimentarius Commission, FAO, Rome.

CODEX (2003) Codex Alimentarius Code of Practice. General Principles of Food Hygiene. CAC/RCP 1-1969. Revision 4. Codex Alimentarius Commission, Rome..

CODEX (2005) Codex Alimentarius Commission Procedural Manual. ISBN:978-92-5-130341-2 Issued by the Secretariat of the Joint FAO/WHO Food Standards Programme 26th Edition. FAO, Rome.

CODEX (2007) Codex Alimentarius Guideline. Guidelines on the application of general principles of food hygiene to the control of *Listeria monocytogenes* in foods. CAC/GL 61 – 2007. Codex Alimentarius Commission, Rome.

CODEX (2013a) Codex Alimentarius Guideline. Principles and Guidelines for the Establishment and Application of Microbiological Criteria Related to Foods. CAC/GL 21 – 1997. Revised and renamed 2013. Codex Alimentarius Commission, Rome.

CODEX (2013b) Codex Alimentarius Commission Guideline. Guidelines for the validation of food safety control measures. CAC/GL 69 – 2008 Amended 2013. Codex Alimentarius Commission, FAO, Rome.

DEPARTMENT OF HEALTH (1994) Guidelines for the safe production of heat preserved foods. ISBN:011321801X London, HMSO.

DOYLE, M.P. (Ed). (1989). Foodborne Bacterial Pathogens. Marcel-Dekker, Inc., New York.

DUIZER, E. AND KOOPMANS, M. (2009) Gastroenteritis viruses. Chp. 32 p.1161-1192 In 'Foodborne Pathogens', Eds. Clive de W. Blackburn & Peter J. McClure, 2nd Edition, ISBN: 978-1-84569-362-6 Woodhead Publishing, UK.

EUROPEAN COMMISSION (1991) Council Directive 91/492/EEC laying down the health conditions for the production and placing on the market of live bivalve molluscs. Official Journal of the European Communities 24.9.91 No. L268/1.

EUROPEAN COMMISSION (2005) Regulation (EC) No. 2073/2005 of 15 November 2005 on microbiological criteria for foodstuffs. Official Journal of the European Union L338/1, 22.12.2005.

EUROPEAN COMMISSION (2005) Regulation (EC) No. 2074/2005 laying down implementing measures for certain products under regulation (EC) No. 853/2004 of the European Parliament and of the Council and for the organisation of official controls under Regulation (EC) No. 854/2004 of the European Parliament and of the Council, derogating from Regulation (EC) No. 852/2004 of the European Parliament and of the Council and amending

Regulations (EC) No 853/2004 and (EC) No 854/2004. Official Journal of the European Union L338/27, 22.12.2005.

EUROPEAN COMMISSION (2006) Regulation (EC) No. 1881/2006 of Commission Regulation (EC) No. 1881/2006 of 19 December 2006 setting maximum levels for certain contaminants in foodstuffs. Official Journal of the European Union L364/5, 20.12.2006.

EUROPEAN COMMISSION (2004) Regulation (EC) No. 853/2004 of the European Parliament and of the Council of 29 April 2004 on the hygiene of foodstuffs. Official Journal of the European Union L139/1, 30.4.2004.

EUROPEAN FOOD SAFETY AUTHORITY (EFSA) (2004). Opinion of the Scientific Panel on Biological Hazards on *Bacillus cereus* and other *Bacillus* spp. in foodstuffs. The EFSA Journal, 175, 1-49. doi:10.2903/j.efsa.2005.175

EUROPEAN FOOD SAFETY AUTHORITY (EFSA) (2005). Opinion of the Scientific Panel on Biological Hazards on the request from the Commission related to *Campylobacter* in animals and foodstuffs. The EFSA Journal, 173, 1-10. doi: 10.2903/j.efsa.2005.173

EUROPEAN FOOD SAFETY AUTHORITY (EFSA) (2007). Opinion of the Scientific Panel on Biological Hazards on Microbiological criteria and targets based on risk analysis. The EFSA Journal 462, 1-29. doi: 10.2903/j.efsa.2007.462

EUROPEAN FOOD SAFETY AUTHORITY (EFSA) (2016) Risks for public health related to the presence of *Bacillus cereus* and other *Bacillus* spp. including *Bacillus thuringiensis* in foodstuffs. The EFSA Journal, 14(7), 4524. doi:10.2903/j.efsa.2016.4524

GMA (2009): GMA, Grocery Manufacturers Association (2009) Control of *Salmonella* in low moisture foods. Accessed June 2019. Available at: https://www.gmaonline.org/downloads/technical-guidance-and-tools/*Salmonella*ControlGuidance.pdf

HEALTH PROTECTION AGENCY (HPA) (2000) Guidelines for the microbiological quality of some ready-to-eat foods sampled at the point of sale. Communicable Disease and Public Health, 3 (3) 163-167.

HEALTH PROTECTION AGENCY (HPA) (2009) Guidelines for Assessing the Microbiological Safety of Ready To Eat Foods Placed on the Market. London:

Health Protection Agency. November,2009. Accessed June 2019. Available at:
https://webarchive.nationalarchives.gov.uk/20140714111812/http://www.hpa.org.uk/webc/HPAwebFile/HPAweb_C/1259151921557

HERSOM, A.C. & HULLAND, E.D. (1980). Canned Foods - Thermal Processing and Microbiology. 7th Edition. ISBN:0443021228 Churchill Livingstone, New York, Edinburgh.

HUSS, H. H., ABABOUCH, L., & GRAM, L. (2003). Assessment and management of seafood safety and quality. FAO Fisheries Technical Paper 444, i-vii, 1-230. ISBN:92-5-104954-8 FAO, Rome.

IID Study (2000) A report Of The Study In Infectious Intestinal Disease In England. ISBN:0-11-322308-0 UK Food Standards Agency London: HMSO. Accessed June 2019. Available at:
https://www.food.gov.uk/sites/default/files/media/document/iid1_study_final_report.pdf .

IID Study (2012) The Second Study Of Infectious Intestinal Disease In The Community (IID2 Study), FSA Project Number: B18021; UK Food Standards Agency, November 2012. Accessed June 2019. Available at
https://www.food.gov.uk/sites/default/files/media/document/711-1-1393_IID2_FINAL_REPORT.pdf

ICMSF (1978). International Commission on Microbiological Specifications for Foods. Microorganisms in Foods Book 1. Their significance and methods of enumeration. 2nd Edition ISBN:0802022936. University of Toronto Press, Toronto.

ICMSF (1980). International Commission on Microbiological Specifications for Foods. Microorganisms in Foods Book 3. Microbial Ecology of Foods Vol. 1. Factors Affecting Life and Death of Microorganisms. ISBN:0123635012 Academic Press, London.

ICMSF (1986). International Commission on Microbiological Specifications for Foods. Microorganisms in Foods Book 2. Sampling for Microbiological Analysis: Principles and Specific Applications. 2nd Edition. ISBN:0802056938 University of Toronto Press, Toronto.

ICMSF (1988). International Commission on Microbiological Specifications for Foods Microorganisms in Foods Book 4. Application of the hazard analysis critical control point (HACCP) system to ensure microbiological safety and quality. ISBN:0632021810 Blackwell Scientific Publications, Oxford.

ICMSF (1996). International Commission on Microbiological Specifications for Foods Microorganisms in Food Book 5. Microbiological Characteristics of Food Pathogens. ISBN:041247350X Blackie Academic & Professional, London.

ICMSF (2011) International Commission on Microbiological Specifications for Foods Microorganisms in Foods Book 8. Use of Data for Assessing Process Control and Products Acceptance. ISBN:978-1-4419-9373-1 Springer-Verlag, New York.

ICMSF (2018) International Commission on Microbiological Specifications for Foods Microorganisms in Foods Book 7. Microbiological Testing in Food Safety Management. 2nd Edition. ISBN:0306472627 Springer-Verlag, New York.

INTERNATIONAL STANDARDS ORGANISATION (ISO) (2001) ISO 17410:2001 Microbiology of food and animal feeding stuffs -- Horizontal method for the enumeration of psychrotrophic microorganisms. International Organization for Standardization, Geneva.

INTERNATIONAL STANDARDS ORGANISATION (ISO) (2005) ISO 17025:2005 General requirements for the competence of testing and calibration laboratories. International Organization for Standardization, Geneva.

INTERNATIONAL STANDARDS ORGANISATION (ISO) (2013) ISO 4833-1:2013 Microbiology of the food chain -- Horizontal method for the enumeration of microorganisms – Part 1: colony count at 30 degrees C by the pour plate technique. International Organization for Standardization, Geneva.

INTERNATIONAL STANDARDS ORGANISATION (ISO) (2016) ISO 18744:2016 Microbiology of the food chain -- Detection and enumeration of *Cryptosporidium* and *Giardia* in fresh leafy green vegetables and berry fruits. International Organization for Standardization, Geneva.

INTERNATIONAL STANDARDS ORGANISATION (ISO) (2016) ISO 16140-1:2016 Microbiology of the food chain -- Method validation -- Part 1: Vocabulary. International Organization for Standardization, Geneva.

INTERNATIONAL STANDARDS ORGANISATION (ISO) (2016) ISO 16140-2:2016 Microbiology of the food chain -- Method validation -- Part 2: protocol for the validation of alternative (proprietary) methods against a reference method. International Organization for Standardization, Geneva.

INTERNATIONAL STANDARDS ORGANISATION (ISO) (2017a) ISO 15216-1:2017 Microbiology of the food chain -- Horizontal method for the determination of hepatitis A virus and norovirus using real-time RT-PCR-Part 1: Method of quantification. International Organization for Standardization, Geneva.

INTERNATIONAL STANDARDS ORGANISATION (ISO) (2017b) ISO 18465:2017 Microbiology of the food chain -- Quantitative determination of emetic toxin (cereulide) using LC-MS/MS. International Organization for Standardization, Geneva.

INTERNATIONAL STANDARDS ORGANISATION (ISO) (2017c) ISO 21872-1:2017 Microbiology of the food chain -- Horizontal method for the determination of *Vibrio* spp. -- Part 1: Detection of potentially enteropathogenic *Vibrio parahaemolyticus*, *Vibrio cholera* and *Vibrio vulnificus*. International Organization for Standardization, Geneva.

INTERNATIONAL STANDARDS ORGANISATION (ISO) (2017d) ISO 22964:2017 Microbiology of the food chain – Horizontal method for the detection of *Cronobacter* spp. International Organization for Standardization, Geneva.

INTERNATIONAL STANDARDS ORGANISATION (ISO) (2017e) ISO 6579-1:2017 Microbiology of the food chain -- Horizontal method for the detection enumeration and serotyping of *Salmonella* -- Part 1 : detection of *Salmonella* spp. International Organization for Standardization, Geneva.

INTERNATIONAL STANDARDS ORGANISATION (ISO) (2017f) ISO/TS 15216-2:2019 Microbiology of food and animal feed -- Horizontal method for the immunoenzymatic detection of staphylococcal enterotoxins in foodstuffs. International Organization for Standardization, Geneva.

INTERNATIONAL STANDARDS ORGANISATION (ISO) (2017g) ISO/TS 15216-2:2019 Microbiology of food and animal feed -- Horizontal method for the detection and enumeration of Enterobacteriaceae -- Part 2: colony count technique. International Organization for Standardization, Geneva.

INTERNATIONAL STANDARDS ORGANISATION (ISO) (2019) ISO/TS 15216-2:2019 Microbiology of food and animal feed -- Horizontal method for determination of hepatitis A virus and norovirus in food using real-time RT-PCR -- Part 2: Method for qualitative detection. International Organization for Standardization, Geneva.

INTERNATIONAL LIFE SCIENCES INSTITUTE (ILSI) (2004) A simple guide to understanding and applying the hazard analysis critical control point concept, 3rd Edition. ISBN:1-57881-179-1 International Life Sciences Institute Europe, Belgium. Accessed June 2019. Available at: https://ilsi.eu/publication/a-simple-guide-to-understanding-and-applying-the-hazard-analysis-critical-control-point-concept/

INSTITUTE OF FOOD SCIENCE AND TECHNOLOGY (2018). Food and Drink - Good Manufacturing Practice: A Guide to its Responsible Management. 7th Edition. ISBN-10:1119388449 Wiley-Blackwell, London.

JARVIS, B. (2007) On the compositing of samples for qualitative microbiological testing. Letters in Applied Microbiology, 45, 592-598.

JENSEN, I. & MOIR, C.J. (2003) *Bacillus cereus* and other *Bacillus* species. Chp. 14. In 'Foodborne microorganisms of public health significance.' Ed. Hocking, A.D. 6th Edition. ISBN:0957890710 Australian Institute of Food Science and Technology (NSW Branch), Sydney, p 445-478.

JORDAN, S. & LAVENDER, B. (20012) Catalogue of Rapid Microbiological Methods, Review No. 1, 7th Edition ISBN:978-0-907503-72-9 Campden & Chorleywood Food Research Association, Chipping Campden, UK.

JUNEIA, V.K. & SOFOS, J.N. (2010) Pathogens and Toxins in Foods: Challenges and Interventions, American Society for Microbiology, Print ISBN: 9781555814595, e-ISBN: 9781555815936, DOI: 10.1128/9781555815936

KRAMER, J.M. & GILBERT, R.J. (1989) *Bacillus cereus* and other *Bacillus* species. In 'Foodborne Bacterial Pathogens', Ed. Doyle, M.P. ISBN:0-8247-7866-9 Marcel Dekker, Inc. New York, USA. p.21-50

McMEEKIN, T.A. (Ed.) (2003) Detecting pathogens in food. ISBN:9781855736702 Woodhead Publishing Ltd, Cambridge.

MOSSEL, D.A.A., CORRY, J.E.L., STRUIJK, C.B. & BAIRD, R.M. (1995). Essentials of the Microbiology of Foods. A textbook for advanced studies. ISBN-10:0471930369 John Wiley & Sons, Chichester.

MURRAY, J. G. (1969) An approach to bacteriological standards. Journal of Applied Bacteriology 32, 123-135.

NATIONAL RESEARCH COUNCIL (1985) National Research Council (US) Subcommittee on Microbiological Criteria, Committee on Food Protection, Food and Nutrition Board. An Evaluation of the Role of Microbiological Criteria for Foods and Food Ingredients. ISBN-10:0-309-03497-3 National Academies Press, Washington D. C.

OH, E., McMULLEN, L. & JEON, B (2015) High prevalence of hyper-aero tolerant *Campylobacter jejuni* in retail poultry with potential implication in human infection. Frontiers in Microbiology 6, article 1263, doi: 10.3389/fmicb.2015.01263.

OH, E., McMULLEN, L., CHUI, L. & JEON, B. (2017) Differential survival of hyper-aerotolerant *Campylobacter jejuni* under different gas conditions. Frontiers in Microbiology, 8, article 954, doi: 10.3389/fmicb.2017.00954.

ROBERTS, D. & GREENWOOD, M. (2003) Practical Food Microbiology. 3rd Edition. ISBN:9781405100755 Blackwell Publishing, Oxford, UK.

SHAPTON, D. A. & SHAPTON, N. F. (Eds.) (1991) Principles and Practices for the Safe Processing of Foods. ISBN:9781855733626 Butterworth-Heinemann Ltd., Oxford.

WOOD, R. & CORRY, J.E.L. (2003) Quality assurance of laboratory performance. In 'Detecting Pathogens in Food', Ed. McMeekin, T.A. ISBN:9781855736702 Woodhead Publishing Ltd, Cambridge, Chp. 5, pp. 93-119.

ABOUT IFST

IFST is the UK's leading professional body for those involved in all aspects of food science and technology. We are an internationally respected independent membership body, supporting food professionals through knowledge sharing and professional recognition.

Our core aim is the advancement of food science and technology based on impartial science and knowledge sharing.

Our membership comprises individuals from a wide range of backgrounds, from students to experts, working across a wide range of disciplines within the sector. Our main areas of activity are:

- **Setting and Recognising Professional Standards**
  Our professional registers are widely recognised and valued throughout the food sector as the benchmark for skills and expertise for food professionals.

- **Delivering an Independent Voice**
  As a charity and professional body that is independent of governments, industry and other organisations, IFST is in a prime position to provide impartial, science-based resources.

- **Providing an Accessible Knowledge Base**
  We facilitiate the exchange of knowledge amongst food professionals, ensuring it is evidence-based, understandable and accessible.

- **Supporting the Future of Food**
  We actively support and encourage the future generation of food scientists and technologists through a wide range of aactivities and events.

For further information contact:
Institute of Food Science & Technology
5 Cambridge Court, 210 Shepherd's Bush Road, London W6 7NJ, UK
Email: info@ifst.org

Institute of Food Science and Technology is a registered charity, number 1121681 and a company limited by guarantee, registered in England and Wales, number 00930776.
Registered address as above.

Published By

Institute of Food Science & Technology

5 Cambridge Court,

210 Shepherd's Bush Road

London W6 7NJ, UK

www.ifst.org

Lightning Source UK Ltd.
Milton Keynes UK
UKHW021308221020
372041UK00005B/1059

9 781916 343801